AF327427

INNOVATIVE SITE REMEDIATION TECHNOLOGY: DESIGN AND APPLICATION

STABILIZATION/ SOLIDIFICATION

One of a Seven-Volume Series

Prepared by WASTECH®, a multiorganization cooperative project managed by the American Academy of Environmental Engineers® with grant assistance from the U.S. Environmental Protection Agency, the U.S. Department of Defense, and the U.S. Department of Energy.

The following organizations participated in the preparation and review of this volume:

 **Air & Waste Management Association**
P.O. Box 2861
Pittsburgh, PA 15230

 **American Society of Mechanical Engineers**
345 East 47th Street
New York, NY 10017

 **American Academy of Environmental Engineers®**
130 Holiday Court, Suite 100
Annapolis, MD 21401

 **Hazardous Waste Action Coalition**
1015 15th Street, N.W., Suite 802
Washington, D.C. 20005

 **American Institute of Chemical Engineers**
345 East 47th Street
New York, NY 10017

 **Soil Science Society of America**
677 South Segoe Road
Madison, WI 53711

 **American Society of Civil Engineers**
345 East 47th Street
New York, NY 10017

 **Water Environment Federation**
601 Wythe Street
Alexandria, VA 22314

Monograph Principal Authors:

Paul D. Kalb, *Chair*
Jesse R. Conner
John L. Mayberry, P.E.

Bhavesh R. Patel
Joseph M. Perez, Jr.
Russell L. Treat

Series Editor
William C. Anderson, P.E., DEE

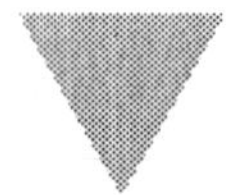

Library of Congress Cataloging in Publication Data

Innovative site remediation technology: design and application.

      p. cm.

"Principle authors: Leo Weitzman, Irvin A. Jefcoat, Byung R. Kim"--V.2, p. iii.

"Prepared by WASTECH."

Includes bibliographic references.

Contents: --[2] Chemical treatment

1. Soil remediation--Technological innovations. 2. Hazardous waste site remediation--Technological innovations. I. Weitzman, Leo. II. Jefcoat, Irvin A. (Irvin Atly) III. Kim, B.R. IV. WASTECH (Project)

TD878.I55   1997

628.5'5--dc21   97-14812

                                      CIP

ISBN 1-883767-17-2 (v. 1)   ISBN 1-883767-21-0 (v. 5)

ISBN 1-883767-18-0 (v. 2)   ISBN 1-883767-22-9 (v. 6)

ISBN 1-883767-19-9 (v. 3)   ISBN 1-883767-23-7 (v. 7)

ISBN 1-883767-20-2 (v. 4)

Printed in the United States of America.

WASTECH and the American Academy of Environmental Engineers are trademarks of the American Academy of Environmental Engineers registered with the U.S. Patent and Trademark Office.

Cover design by William C. Anderson. Cover photos depict remediation of the Scovill Brass Factory, Waterbury, Connecticut, recipient of the 1997 Excellence in Environmental Engineering Grand Prize award for Operations/Management.

CONTRIBUTORS

PRINCIPAL AUTHORS

Paul D. Kalb, *Task Group Chair*
Brookhaven National Laboratory

Jesse R. Conner
Conner Technologies, Inc.

Bhavesh R. Patel
U.S. Department of Energy

John L. Mayberry, P.E.
SAIC

Joseph M. Perez, Jr.
Battelle Pacific Northwest

Russell L. Treat
MACTEC

The authors gratefully acknowledge David Eaton and Margaret Knecht, Lockheed Martin Idaho Technologies, for their assistance in reviewing and editing sections in this monograph on regulatory issues.

REVIEWERS

The panel that reviewed the monograph under the auspices of the Project Steering Committee was composed of:

Joseph F. Lagnese, Jr., P.E., DEE, *Chair*
Allison Park, Pennsylvania

Roger Olsen, Ph.D.
Camp Dresser and McKee

Ed Barth, P.E.
National Risk Management
 Research Laboratory
USEPA

Tim Oppelt
National Risk Management
 Research Laboratory
USEPA

T. Michael Gilliam, P.E.
Oak Ridge National Laboratory

William C. Webster
Webster and Associates

STEERING COMMITTEE

This monograph was prepared under the supervision of the WASTECH® Steering Committee. The manuscript for the monograph was written by a task group of experts in chemical treatment and was, in turn, subjected to two peer reviews. One review was conducted under the auspices of the Steering Committee and the second by professional and technical organizations having substantial interest in the subject.

Frederick G. Pohland, Ph.D., P.E., DEE *Chair*
Weidlein Professor of Environmental
  Engineering
University of Pittsburgh

Richard A. Conway, P.E., DEE, *Vice Chair*
Senior Corporate Fellow
Union Carbide Corporation

William C. Anderson, P.E., DEE
Project Manager
Executive Director
American Academy of Environmental
  Engineers

Colonel Frederick Boecher
U.S. Army Environmental Center
Representing American Society of Civil
  Engineers

Clyde J. Dial, P.E., DEE
Manager, Cincinnati Office
SAIC
Representing American Academy of
  Environmental Engineers

Timothy B. Holbrook, P.E.
Engineering Manager
Camp Dresser & McKee, Incorporated
Representing Air & Waste Management
  Association

Joseph F. Lagnese, Jr., P.E., DEE
Private Consultant
Representing Water Environment Federation

Peter B. Lederman, Ph.D., P.E., DEE, P.P.
Center for Env. Engineering & Science
New Jersey Institute of Technology
Representing American Institute of Chemical
  Engineers

George O'Connor, Ph.D.
University of Florida
Representing Soil Science Society of America

George Pierce, Ph.D.
Manager, Bioremediation Technology Dev.
American Cyanamid Company
Representing the Society of Industrial
  Microbiology

Peter W. Tunnicliffe, P.E., DEE
Senior Vice President
Camp Dresser & McKee, Incorporated
Representing Hazardous Waste Action
  Coalition

Charles O. Velzy, P.E., DEE
Private Consultant
Representing, American Society of
  Mechanical Engineers

Calvin H. Ward, Ph.D.
Foyt Family Chair of Engineering
Rice University
At-large representative

Walter J. Weber, Jr., Ph.D., P.E., DEE
Gordon Fair and Earnest Boyce Distinguished
  Professor
University of Michigan
Representing Hazardous Waste Research Centers

FEDERAL REPRESENTATION

Walter W. Kovalick, Jr., Ph.D.
Director, Technology Innovation Office
U.S. Environmental Protection Agency

George Kamp
Cape Martin Energy Systems
U.S. Department of Energy

Jeffrey Marqusee
Office of the Under Secretary of Defense
U.S. Department of Defense

Timothy Oppelt
Director, Risk Reduction Engineering
  Laboratory
U.S. Environmental Protection Agency

REVIEWING ORGANIZATIONS

The following organizations contributed to the monograph's review and acceptance by the professional community. The review process employed by each organization is described in its acceptance statement. Individual reviewers are, or are not, listed according to the instructions of each organization.

Air & Waste Management Association

The Air & Waste Management Association is a nonprofit technical and educational organization with more than 14,000 members in more than fifty countries. Founded in 1907, the Association provides a neutral forum where all viewpoints of an environmental management issue (technical, scientific, economic, social, political, and public health) receive equal consideration.

Qualified reviewers were recruited from the Waste Group of the Technical Council. It was determined that the monograph is technically sound and publication is endorsed.

The reviewers were:

Terry Alexander, P.E., DEE, CIH

    University of Michigan

Tim Holbrook, P.E., DEE

    Camp Dresser & McKee

Charles Wilk

    Portland Cement Association

American Institute of Chemical Engineers

The Environmental Division of the American Institute of Chemical Engineers has enlisted its members to review the monograph. Based on that review the Environmental Division endorses the publication of the monograph.

American Society of Civil Engineers

The American Society of Civil Engineers, established in 1852, is the premier civil engineering association in the world with 124,000 members. Qualified reviewers were recruited from its Environmental Engineering Division.

ASCE has reviewed this manual and believes that significant information of value is provided. Many of the issues addressed, and the resulting conclusions, have been evaluated based on satisfying current regulatory requirements. However, the long-term stability of solidified soils containing high levels of organics, and potential limitations and deficiencies of current testing methods, must be evaluated in more detail as these technologies are implemented and monitored.

The reviewers included:

Richard Reis, P.E.

    Fluor Daniel GTI

    Seattle, WA

G. Fred Lee, Ph.D., P.E., DEE

    G. Fred Lee & Associates

    El Macero, CA

American Society of Mechanical Engineers

Founded in 1880, the American Society of Mechanical Engineers (ASME) is a nonprofit educational and technical organization, having at the date of publication of this document approximately 116,400 members, including 19,200 students. Members

"

work in industry, government, academia, and consulting. The Society has thirty-seven technical divisions, four institutes, and three interdisciplinary programs which conduct more than thirty national and international conferences each year.

This document was reviewed by volunteer members of the Research Committee on Industrial and Municipal Waste, each with technical expertise and interest in the field covered by the document. Although, as indicated on the reverse of the title page of this document, neither ASME nor any of its Divisions or Committees endorses or recommends, or makes any representation or warranty with respect to, this document, those Divisions and Committees which conducted a review believe, based upon such review, that this document and findings expressed are technically sound.

Hazardous Waste Action Coalition

The Hazardous Waste Action Coalition (HWAC) is the premier business trade group serving and representing the leading engineering and science firms in the environmental management and remediation industry. HWAC's mission is to serve and promote the interests of engineering and science firms practicing in multi-media environment management and remediation. Qualified reviewers were recruited from HWAC's Technical Practices Committee. HWAC is pleased to endorse the monograph as technically sound.

The lead reviewer was:

James D. Knauss, Ph.D.

President, Shield Environmental

Lexington, KY

Soil Science Society of America

The Soil Science Society of America, headquartered in Madison, Wisconsin, is home to more than 5,300 professionals dedicated to the advancement of soil science. Established in 1936, SSSA has members in more than 100 countries. The Society is composed of eleven divisions, covering subjects from the basic sciences of physics and chemistry through soils in relation to crop production, environmental quality, ecosystem sustainability, waste management and recycling, bioremediation, and wise land use.

Members of SSSA have reviewed the monograph and have determined that it is acceptable for publication.

The lead reviewer was:

Chet Francis

Oak Ridge National Laboratory

Water Environment Federation

The Water Environment Federation is a nonprofit, educational organization composed of member and affiliated associations throughout the world. Since 1928, the Federation has represented water quality specialists including engineers, scientists, government officials, industrial and municipal treatment plant operators, chemists, students, academic and equipment manufacturers, and distributors.

Qualified reviewers were recruited from the Federation's Hazardous Wastes Committee and from the general membership. It has been determined that the document is technically sound and publication is endorsed.

The lead reviewer was:

Robert C. Williams, P.E., DEE

ACKNOWLEDGMENTS

The WASTECH® project was conducted under a cooperative agreement between the American Academy of Environmental Engineers® and the Office of Solid Waste and Emergency Response, U.S. Environmental Protection Agency. The substantial assistance of the staff of the Technology Innovation Office was invaluable.

Financial support was provided by the U.S. Environmental Protection Agency, Department of Defense, Department of Energy, and the American Academy of Environmental Engineers®.

This multiorganization effort involving a large number of diverse professionals and substantial effort in coordinating meetings, facilitating communications, and editing and preparing multiple drafts was made possible by a dedicated staff provided by the American Academy of Environmental Engineers® consisting of:

William C. Anderson, P.E., DEE
Project Manager & Editor

John M. Buterbaugh
Assistant Project Manager & Managing Editor

Karen Tiemens
Editor

Catherine L. Schultz
Yolanda Y. Moulden
Project Staff Production

J. Sammi Olmo
I. Patricia Violette
Project Staff Assistants

TABLE OF CONTENTS

LIST OF FIGURES

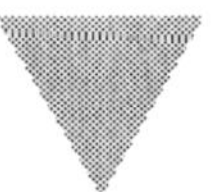

Figure	Title	Page

1

INTRODUCTION

This monograph, covering the *design, applications, and implementation* of Stabilization/Solidification, is one of a series of seven on innovative site and waste remediation technologies. This series of seven monographs by the American Academy of Environmental Engineers® was preceded by eight volumes published in 1994 and 1995 covering the description, evaluation, and limitations of the processes. The entire project is the culmination of a multiorganization effort involving more than 100 experts. It provides the experienced, practicing professional guidance on the innovative processes considered ready for full-scale application. Other monographs in this design and application series and the companion series address bioremediation; chemical treatment; liquid extraction: soil washing, soil flushing, and solvent/chemical extraction; thermal desorption; thermal destruction; and vapor extraction and air sparging.

1.1 Stabilization/Solidification

Stabilization and Solidification are generic names applied to a wide range of discrete technologies. These technologies, i.e., Stabilization and Solidification, are closely related in that both use chemical, physical, and/or thermal processes to reduce potential adverse impacts on the environment from the disposal of radioactive, hazardous, and mixed waste. But, they are distinct technologies.

Stabilization refers to techniques that reduce the hazard potential of a waste by converting the contaminants into less soluble, mobile, or toxic forms. The physical nature and handling characteristics of the waste are not necessarily changed by stabilization.

Solidification refers to techniques that encapsulate the waste, forming a solid material. The product of solidification, often known as the waste form, may be a monolithic block, a clay-like material, a granular particulate, or some other physical form commonly considered "solid." Solidification as applied to fine waste particles is termed *microencapsulation* and that which applies to a large block or container of wastes is termed *macroencapsulation*. Solidification can be accomplished by a chemical reaction between the waste and solidifying reagents or by mechanical processes. Contaminant migration is often restricted by decreasing the surface area exposed to leaching and/or by coating the wastes with low-permeability materials.

Each of the proven innovative technologies which are included within the general categories of stabilization and solidification are discussed in this monograph.

They can be grouped into three broad categories: (1) aqueous stabilization/ solidification, (2) polymer stabilization/solidification, and (3) vitrification.

1.2 Development of the Monograph

1.2.1 Background

Acting upon its commitment to develop innovative treatment technologies for the remediation of hazardous waste sites and contaminated soils and groundwater, the U.S. Environmental Protection Agency (US EPA) established the Technology Innovation Office (TIO) in the Office of Solid Waste and Emergency Response in March, 1990. The mission assigned TIO was to foster greater use of innovative technologies.

In October of that same year, TIO, in conjunction with the National Advisory Council on Environmental Policy and Technology (NACEPT), convened a workshop for representatives of consulting engineering firms, professional societies, research organizations, and state agencies involved in remediation. The workshop focused on defining the barriers that were impeding the application of innovative technologies in site remediation projects. One of the major impediments identified was the lack of reliable data on the performance, design parameters, and costs of innovative processes.

The need for reliable information led TIO to approach the American Academy of Environmental Engineers®. The Academy is a long-standing, multidisciplinary environmental engineering professional society with wide-ranging affiliations with the remediation and waste treatment professional communities. By June, 1991, an agreement in principle (later formalized as a Cooperative Agreement) was reached providing for the Academy to manage a project to develop monographs describing the state of available innovative remediation technologies. Financial support was provided by the US EPA, U.S. Department of Defense (DoD), U.S. Department of Energy (DOE), and the Academy. The goal of both TIO and the Academy was to develop monographs providing reliable data that would be broadly recognized and accepted by the professional community, thereby eliminating or at least minimizing this impediment to the use of innovative technologies.

The Academy's strategy for achieving the goal was founded on a multiorganization effort, WASTECH® (pronounced Waste Tech), which joined in partnership the Air and Waste Management Association, the American Institute of Chemical Engineers, the American Society of Civil Engineers, the American Society of Mechanical Engineers, the Hazardous Waste Action Coalition, the Society for Industrial Microbiology, the Soil Science Society of America, and the Water Environment Federation, together with the Academy, US EPA, DoD, and DOE. A Steering Committee composed of highly respected representatives of these organizations having expertise in remediation technology formulated the specific project objectives and process for developing the monographs (see page iv for a listing of Steering Committee members).

By the end of 1991, the Steering Committee had organized the Project. Preparation of the initial monographs began in earnest in January, 1992, and the original eight monographs were published during the period of November, 1993, through April, 1995. In Spring of 1995, based upon the receptivity of the industry and others of the original monographs, it was determined that a companion set, emphazing the design and applications of the technologies, should be prepared as well. Task Groups were identified during the latter months of 1995 and work commenced on this second series.

1.2.2 Process

For each of the series, the Steering Committee decided upon the technologies, or technological areas, to be covered by each monograph, the monographs' general scope, and the process for their development and appointed a task group composed of five or more experts to write a manuscript for each monograph. The task groups were appointed with a view to balancing the interests of the groups principally concerned with the application of innovative site and waste remediation technologies — industry, consulting engineers, research, academe, and government.

The Steering Committee called upon the task groups to examine and analyze all pertinent information available, within the Project's financial and time constraints. This included, but was not limited to, the comprehensive data on remediation technologies compiled by US EPA, the store of information possessed by the task groups' members, that of other experts willing to voluntarily contribute their knowledge, and information supplied by process vendors.

To develop broad, consensus-based monographs, the Steering Committee prescribed a twofold peer review of the first drafts. One review was conducted by the Steering Committee itself, employing panels consisting of two members of the Committee supplemented by at least four other experts (See *Reviewers,* page iii, for the panel that reviewed this monograph). Simultaneous with the Steering Committee's review, each of the professional and technical organizations represented in the Project reviewed those monographs addressing technologies in which it has substantial interest and competence.

Comments resulting from both reviews were considered by the Task Group, appropriate adjustments were made, and a second draft published. The second draft was accepted by the Steering Committee and participating organizations. The statements of the organizations that formally reviewed this monograph are presented under *Reviewing Organizations* on page v.

1.3 Purpose

The purpose of this monograph is to further the use of innovative stabilization/solidification site remediation and waste processing technologies, that is, technologies not commonly applied, where their use can provide better, more cost-effective performance than conventional methods. To this end, the monograph documents the current state of stabilization/solidification technology.

1.4 Objectives

The monograph's principal objective is to furnish guidance for experienced, practicing professionals and project managers charged with site remediation responsibility. The monograph, and its companion monograph, are intended, therefore, not to be prescriptive, but supportive. It is intended to aid experienced professionals in applying their judgment in deciding whether and how to apply the technologies addressed under the particular circumstances confronted.

In addition, the monograph is intended to inform regulatory agency personnel and the public about the conditions under which the processes it addresses are potentially applicable.

1.5 Scope

The monograph addresses innovative stabilization/solidification technologies that have been sufficiently developed so that they can be used in full-scale applications. It addresses all aspects of the technologies for which sufficient data were available to the Stabilization/Solidification Task Group to briefly review the technologies and discuss their design and applications. Actual case studies were reviewed and included, as appropriate.

The monograph's primary focus is site remediation and waste treatment. To the extent the information provided can also be applied elsewhere, it will provide the profession and users this additional benefit.

Application of site remediation and waste treatment technology is site-specific and involves consideration of a number of matters besides alternative technologies. Among them are the following that are addressed only to the extent that they are essential to understand the applications and limitations of the technologies described:

- site investigations and assessments;
- planning, management, specifications, and procurement;
- cost-benefit analyses;
- regulatory requirements; and
- community acceptance of the technology.

1.6 Limitations

The information presented in this monograph has been prepared in accordance with generally recognized engineering principles and practices and is for general information only. This information should not be used without first securing competent advice with respect to its suitability for any general or specific application.

Readers are cautioned that the information presented is that which was generally available during the period when the monograph was prepared. Development of innovative site remediation and waste treatment technologies is ongoing. Accordingly, postpublication information may amplify, alter, or render obsolete the information about the processes addressed.

This monograph is not intended to be and should not be construed as a standard of any of the organizations associated with the WASTECH® Project; nor does reference in this publication to any specific method, product, process, or service constitute or imply an endorsement, recommendation, or warranty thereof.

1.7 Organization

This monograph and others in the series are organized under a similar outline intended to facilitate cross reference among them and comparison of the technologies they address.

Chapter 2, Application Concepts, summarizes the scientific basis, potential applications, and key requirements for each stabilization/solidification technology addressed. Design Development, Chapter 3, provides essential information for those contemplating use of the technologies discussed. Chapter 4, Implementation and Operation, focuses on the procedures commonly used to implement stabilization/solidification technologies and key facets of their operation. An evaluation of Case Histories for each technology is provided in Chapter 5.

2

APPLICATION CONCEPTS

Scientific principles, potential applications, and various treatment train configurations for aqueous, polymer, and vitrification Stabilization/Solidification (S/S) processes are briefly reviewed in this chapter. Additional background information on these technologies can be found in the companion monograph, *Innovative Site Remediation Technology: Stabilization/Solidification* (Colombo et al. 1994) and in the references cited throughout this monograph. Prior to selection of technology(ies) for implementing restoration and waste management operations, site remediation managers and engineers are encouraged to conduct a comprehensive evaluation of potential processes. This evaluation should address issues of applicability, maturity, availability, cost-effectiveness, reliability, ease of operation, performance, health and safety, regulatory compliance, and public acceptance. In addition, laboratory-scale feasibility and/or treatability studies should be conducted prior to full-scale implementation to verify the appropriateness of the technology(ies) selected.

A frequent criticism of hazardous waste treatment operations is that the affected public has little, if any, input into the decisions that ultimately impact them directly; namely site and technology selection, operation, and monitoring. Design engineers and project managers need to take community interests into account and include their input in planning and oversight activities. This may require additional time and resources devoted to educating the public on the issues, options, risks, and costs. However, the importance of well-conceived and well-executed public involvement cannot be overemphasized for the successful siting, construction, and operation of most site remediation projects.

Several of the terms used throughout this monograph, although commonly used in reference to waste treatment and environmental restoration activities, are key to the information in this monograph and are, therefore, reviewed here.

Stabilization — Techniques that reduce the hazard potential of a waste by converting the contaminants into less soluble, mobile, or toxic forms. The physical nature and handling characteristics of the waste are not necessarily changed by stabilization.

Solidification — Techniques that encapsulate the waste, forming a solid material. The product of solidification, often known as the waste form, may be a monolithic block, a clay-like material, a granular particulate, or some other physical form commonly considered "solid." Solidification applied to fine waste particles is termed *microencapsulation* and that which applies to a large block or container of wastes is termed *macroencapsulation*. Solidification can be accomplished by a chemical reaction between the waste and solidifying reagents, or by mechanical processes. Contaminant migration is often restricted by decreasing the surface area exposed to leaching and/or by coating the wastes with a low-permeability material.

Ex-Situ Treatment — The treatment of waste materials in an engineered processing system following removal from their original location. Wastes treated by an ex-situ method are usually disposed at a solid waste landfill, licensed Resource Conservation and Recovery Act (RCRA) hazardous waste landfill, radioactive or mixed waste disposal facility.

In Situ Treatment — The processing of waste materials at the location where they currently reside, i.e., without removal from the ground, tank, settling pond, etc. Wastes treated in situ are usually left in place for final disposal.

Microencapsulation — Thorough and homogeneous mixing of small waste particles with a liquid binder which then solidifies to form a solid, monolithic final waste form. Individual waste particles are coated and surrounded by the solidified binder to provide mechanical integrity and act as a barrier against leaching of contaminants.

Macroencapsulation — Compactly packaging large pieces of waste not suitable for processing by microencapsulation (e.g., debris, large pieces of solid metal) and surrounding the package with a layer of clean binder. The binder forms a "cocoon" around the waste, provides structural support, and helps prevent migration of contaminants.

2.1 Aqueous Stabilization/Solidification

2.1.1 Scientific Principles

The earliest and by far most commonly used S/S technologies are referred to as "aqueous S/S" in this monograph because the chemical reactions occur only in the presence of water. They are commonly referred to as Stabilization/Solidification in other literature. This technology is further subdivided into two process types: stabilization alone and cementitious S/S. For environmental restoration applications, the aqueous S/S process can be conducted either ex-situ or in situ.

In aqueous systems, stabilization and solidification usually occur simultaneously, hence the customary term, stabilization/solidification. However, there are many instances in which stabilization is performed without solidifying. The distinction is quite important technically and commercially in remediation as will be discussed later. While stabilization requires chemicals to convert contaminants into less soluble, mobile, or toxic forms, the cementitious binder in S/S processes usually supplies the required reactants for stabilization. Also, since solidification does not occur in waste treated through stabilization alone, the physical properties of the treated waste are not altered significantly. For this reason, the use of stabilization alone is limited to those wastes for which physical encapsulation is not required. On the other hand, stabilization is often easier to achieve than solidification, especially for in situ treatment.

2.1.1.1 Stabilization

Stabilization chemistry is similar to wastewater treatment but the waste to be treated is usually soil, sludge, or similar material. Stabilization processes can treat wastes containing metals (and occasionally, other inorganic species such as cyanides) or organic contaminants, and often both in the same waste. The chemistry involved is very different for the two contaminant types, and will be discussed separately.

Stabilization of Metals and Inorganic Contaminants. Technical descriptions of various reactions that can be used to stabilize metals are given by Conner (1990), Conner (1997), and Wilson and Clarke (1994). The principles of these techniques are well-known, and are usually associated with cementitious S/S. Examples of some of the more important metal stabilization reactions are:

- pH control;

- oxidation/reduction potential (ORP) control; and

- speciation by chemical reaction, including:

    - carbonate precipitation,

    - sulfide precipitation,

    - silicate precipitation,

    - ion-specific precipitation,

    - complexation,

    - adsorption,

    - chemisorption,

    - passivation,

    - ion exchange,

    - diadochy (crystal lattice substitution),

    - reprecipitation, and

    - coprecipitation.

In addition to metals, other inorganic species amenable to stabilization (Conner 1990) are cyanides, sulfides, and fluoride. Traditionally, these species were destroyed or immobilized in a pretreatment operation, but often treatment can be accomplished simultaneously with metal and/or organic stabilization and even with S/S.

While a number of chemicals can be used for speciation of metals (e.g., carbonates, sulfides, etc.), only a few have been developed as discrete, innovative systems. One reason is that the regulations and testing requirements affecting metal stabilization have not significantly changed in recent years (refer to proposed changes in Chapter 3), so there has been little need for improvement or change in conventional technology. Innovative stabilization processes covered in this monograph are the ProFix™ and phosphate systems. Both of these processes can be used in the S/S modes as well. Phosphate treatment is discussed below, and ProFix™ is described is Section 2.1.1.2 under cementitious S/S.

Phosphate treatment, as described in the companion monograph (Colombo et al. 1994), involves adding chemical compounds containing phosphate, which form complexes with the metal species present in the matrix. The phosphate-metal complexes have low solubility and immobilize the metals over a wide pH range. Alkali can also be added for pH control (Eighmy et al. 1991) to treat metals such as cadmium. The phosphate treatment process was developed primarily for stabilization of lead. Also, phosphates have historically been used in wastewater treatment and as additives in stabilization (Krueger, Chowdhury, and Warner 1991) because of the low solubility of the resulting reaction products (Jowett and Price 1932). For waste stabilization, phosphates have been used primarily for materials, such as contaminated soils and incinerator ashes, which retain their particulate nature after treatment. Phosphates can be used in conjunction with cementitious materials to improve the physical characteristics of the treated waste, if desired, in which case the process is usually considered to be a cementitious S/S process with phosphate as an additive.

Immobilization of Organic Contaminants. The immobilization, or stabilization, of low levels of hazardous organic compounds in soils, sludges, debris, and other wastes has recently received increasing attention. Newly adopted rules and testing protocols (refer to Chapter 3) have made the development of innovative stabilization techniques necessary. Previously, the use of additives, such as activated carbon in S/S systems to immobilize organic constituents, was based on meeting the requirements of the Toxicity Characteristic Leaching Procedure (TCLP) test method (US EPA 1986a). However, with the more recent Total Constituent Analysis (TCA) test method, such additives are often not very effective (Lear and Conner 1991). As a result, a number of other reagent additives have been developed or adapted from other technologies to meet the new requirements. Furthermore, the high alkalinity associated with S/S binders can hinder immobilization of organics, making stabilization alone a more attractive alternative than S/S for these wastes. Innovative processes or reagents for stabilization of organics include:

- *Rubber Particulate.* Conner and Smith (1993) described the results of this new process. The stabilization agent was a specially-prepared, finely-ground rubber material called KAX-50™. This material was found to be very effective with semivolatile and nonvolatile organics; pesticides, herbicides, and polychlorinated biphenyls (PCBs); and certain volatile organics and organometallic compounds. Mixed with other additives in a *proprietary* formulation, KAX-100™, rubber particulate performed well on nearly all hazardous organics tested to date — about 60 compounds. In fact, it was the only compound tested that was effective for practical use in reducing total levels of organics across a broad range, as measured by the TCA method.

- *Organo-Clays.* These processes are discussed in the companion monograph (Colombo et al. 1994) under "Sorption and Surfactant Processes." They are still innovative largely because they are now being used in the new regulatory context mentioned previously. Organo-clays are formed by substituting quaternary ammonium ions for group IA and IIA metal ions in clays, increasing the organophilic property of the clay. This substitution can also increase the interplanar distance between alumina and silica layers, allowing organic compounds to intercalate themselves between the layers. Both of these mechanisms result in stronger bonding of organic molecules to the clay substrate. Several companies, including Silicate Technology Corporation (STC), International Waste Technologies (IWT), Hazcon, and Soliditech, have provided full-scale services using organo-clays.

- *Other Sorbents.* A number of other sorbents can be, and have been, used to immobilize organics in specific remediation projects. They include rice hull ash, coal, and petroleum coke. However, none of these technologies has sufficiently advanced to a full-scale application, and therefore they are not covered in this monograph.

- *Surfactants.* The use of surfactants for stabilization stems from the ability of the surfactant molecule to attract and hold organic contaminants at one end, while the other end is attached to an immobile, solid substrate surface such as soil or cementitious material. An example of such a process, demonstrated at field-scale, is described in Section 4.1 of the companion monograph (Colombo et al. 1994).

2.1.1.2 Cementitious Stabilization/Solidification

Cementitious S/S technologies use inorganic reagents to react with certain waste components; they also react among themselves to form chemically and mechanically stable solids. Cementitious binders and other additives react in a controlled manner to produce a solid matrix. The matrix itself often is, or becomes, a pseudo-mineral. This type of structure is stable and has a rigid, friable structure similar to many soils and rocks. Many inorganic stabilization systems require promoters that are described as "inorganic polymers." In one sense, all cementitious systems could be characterized in this way, but within the narrow definition of polymers, i.e., monomers that react to form larger molecules by a chemical polymerization process, the setting of Portland cement does not qualify as polymerization. Various vendors market variations of the system to address specific wastes or disposal scenarios. Vendors often describe a modification or additive as "innovative," but unless there is a significant difference in the S/S mechanism or in the way that it is applied to waste treatment, it is not covered here. For example, the addition of a simple fixative, such as sulfide ion to immobilize mercury is not considered innovative. On the other hand, the use of a conventional stabilization agent, such as soluble silicate, in a different way to treat a specific waste type might be innovative.

Different processes exhibit different setting and curing reactions. Most of the commercial, cementitious S/S systems, however, solidify by similar reactions which have been thoroughly studied in connection with Portland cement technology used in concrete (Conner 1990). While the pozzolanic reactions of the processes using fly ash and kiln dusts are not identical to those of Portland cement, the general reactions are similar. Sufficient water must be provided to support the hydration reaction of these chemical systems. Conventional inorganic chemical processes that have been used commercially are shown below. The most important systems currently on the market are marked with an asterisk:

- Portland cement-based (major ingredient is cement)*;
- Portland cement/lime;
- Portland cement/clay;
- Portland cement/fly ash*;
- Portland cement/soluble silicate*;

- lime/fly ash*;
- cement or lime kiln dust*; and
- slag.

Cementitious processes that are deemed innovative due to either their chemistry or the way in which they are used, are described below.

***ProFix*™ (*EnviroGuard, Inc., Houston, Texas*).** This product is based on rice hull ash, an amorphous, biogenetic silica (Durham and Henderson 1984). Because of its sorptive and alkali-reactive nature, rice hull ash has some unusual properties. Its sorptive nature is well-known, but its ability to react with alkalies to form soluble silicates is of primary interest here. Under alkaline conditions, the amorphous silica reacts slowly to produce soluble silicates, which can then react with toxic metal ions to form low-solubility metal silicates. The scientific basis for this process is covered in some detail in the companion monograph (Colombo et al. 1994). The process has patents and patents-applied-for in many countries (Conner and Reber 1992).

Cement-Slag Processes. Slag has been incorporated into a number of stabilization processes, along with other reagents, especially at or near slag producers such as steel mills. As with other waste product reagents (fly ash, kiln dusts), slag usage is often not documented in the literature or promoted specifically as a commercial S/S process. It is used in a proprietary manner by waste generators and industries. Vitreous blast furnace slag is produced when molten slag from an iron-producing blast furnace is cooled quickly to minimize crystallization. Granulation, the most common process, produces a product known as "granulated blast furnace slag." Other processes such as pelletization, are also used. Blast furnace slag is a blend of amorphous silicates and alumini silicates of calcium and other bases. The vitrified slag must be ground to cement fineness. Because of the presence of ferrous iron and reduced sulfur compounds, the slag may act as a reducing agent for metal species that are less mobile in the reduced oxidation state, such as chromium. Note that slowly-cooled, crystalline slag (e.g., air-cooled slag, foamed slag) does not exhibit hydraulic cementing reactions.

2.1.1.3 In Situ Stabilization and Stabilization/Solidification

For in situ application, S/S binders and additives are introduced into the contaminated medium (usually sludge or soil) using commonly-available, large-scale excavation, tilling, or drilling equipment specially modified for S/S chemical addition. In situ and ex-situ S/S delivery systems each have advantages and disadvantages. Ex-situ systems provide better control of reagent addition and mixing, at least as of this time, and quality control sampling is easier. It is usually more practical for projects with shallow waste depths and where site access for large equipment is limited. For obvious reasons, ex-situ equipment and methods are used at central waste treatment locations, such as RCRA Subtitle C Treatment, Storage, and Disposal Facilities (TSDFs) and are now considered conventional. Large-scale remedial projects at great depth are more amenable to in situ operation, and in these instances, the cost is usually lower. In situ methods were probably the first to be used for remedial projects, long before RCRA, Land Disposal Restrictions (LDRs), Superfund, and the other legislative and regulatory drivers that created the hazardous waste industry. The methods used were rather crude, mostly mixing the binders with contaminated media using the standard backhoe excavation bucket. With time, various devices for injecting S/S reagents and mixing them into the waste were substituted for the bucket, but the operation remained basically the same.

A recent technique using modified, massive earth drilling and foundation construction equipment has been introduced to allow well-controlled reagent injection and mixing even at great depth. While such mechanical systems are not new, and were suggested for this application long ago (Conner 1990), they have only recently been applied to actual full-scale remediation projects, and are deemed innovative. One such system is shown in schematic in Figure 2.1 and in actual operation in Figure 2.2. A slurry of the S/S reagents is pumped through a drilling assembly consisting of a vertical, hollow bar called a "kelly bar" and a set of hollow auger blades, into the soil or sludge as the assembly is rotated down through it. High torque (with forces up to 41,500 kg • m) is available to produce a well-mixed, treated waste at depths up to 30 m (100 ft) or more. The resulting treated column may reach a diameter up to 4.3 m (14 ft). Subsequent columns are positioned to eventually cover the entire volume of waste to be treated.

Figure 2.1
Schematic Diagram of MecTool™ for Solidification
and Stabilization of Contaminated Soils/Sludges

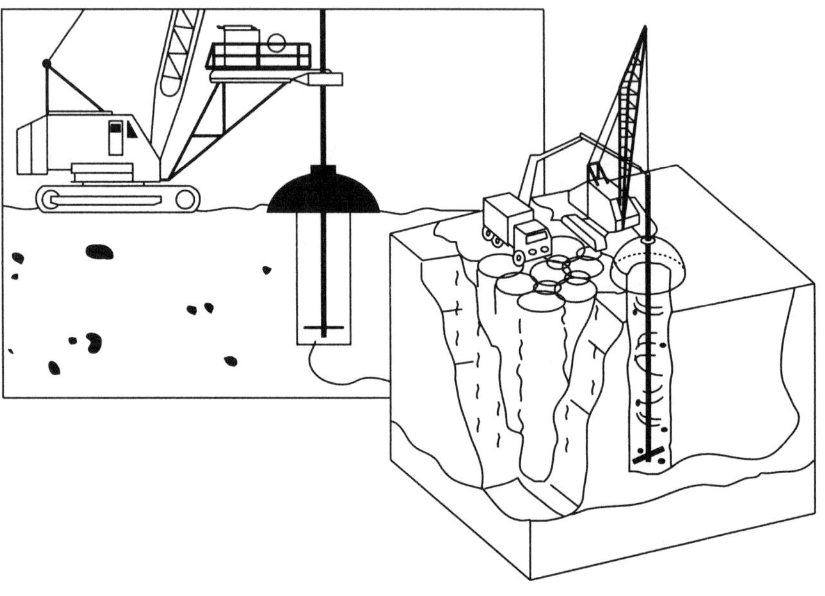

Reproduced courtesy of Millgard Environmental Corp.

A major advantage of this in situ method is its ability to easily and effectively control both volatile and particulate emissions from the site using a hood or shroud over the drilling assembly and the column being stabilized. The control of volatile organic compound (VOC) emissions has become a major issue in remedial work. This is much more difficult at a remedial site than at TSDFs and vastly more expensive. Control of VOCs will likely drive the physical technology toward in situ treatment with equipment that easily collects and treats emissions.

In addition to just S/S processing, in situ methods are available for sequential treatment operations, e.g., pretreatment to oxidize or reduce a constituent followed by solidification. One such multi-step operation that is especially amenable to the drilling type of in situ treatment is stripping volatile organics from the waste prior to metal stabilization. With ex-situ treatment, two different treatment systems must be mobilized to the site and operated; with in situ treatment, the same basic equipment does both more cost-effectively.

Figure 2.2
The MecTool™ System in Operation

Reproduced courtesy of Millgard Environmental Corp.

2.1.2 Potential Applications

2.1.2.1 Stabilization

Soluble silicates or phosphates are used for metal stabilization applications only, not for organic contaminants. In some cases, they can be combined with the various organic immobilization reagents for combined treatment of metals and organics. Care must be taken to ensure combinations of reagents to be used are compatible in the concentration range. Soluble silicate processes, such as ProFix™, can treat a wide range of metals, while phosphates are generally more narrowly used, primarily for lead stabilization. Phosphate treatment produces lead reaction products that are stable and have very low solubility over a wide pH range, making them suitable where the waste form might be exposed to acidic leaching conditions.

Rubber particulate and organo-clays are applicable only for immobilization of organic contaminants and are not normally used for wastes containing metals unless combined with soluble silicates, phosphates, pH control agents, or other metal stabilization reagents. They can also be used in conjunction with other additives. For example, a mixture of rubber particulate and activated carbon might be useful where a wide range of constituents are present or where both TCA and TCLP tests are to be used to evaluate effectiveness. In general, these additives do not interfere with the reactions that occur in cementitious systems and can be mixed in nearly any proportion with cementitious and pozzolanic binders, if specific physical properties are required in the final waste form or if metal stabilization is necessary. The pH of the system can affect the efficiency of the sorbent; this should be ascertained in prior treatability studies for each application.

2.1.2.2 Cementitious Stabilization/Solidification

Many of the various conventional cementitious systems, including additives, are patented or are covered by patents applied for, but most of the generic system types are believed to be in the public domain, at least in the U.S. and Canada (Conner 1990). These processes have been used commercially for solidification of water-based waste liquids, sludges, filter cakes, and contaminated soils. A large body of technical information on leachability, physical properties, and general stability is available. Treatability studies for the particular waste and remedial scenarios being considered are essential for all S/S projects.

For the most part, potential applications are merely extensions of the conventional, well-proven processes that have been used for more than two decades on hazardous and nonhazardous wastes. These extended applications or improvements include decreasing leachability by chemical reaction or microencapsulation, reducing permeability, and improving physical durability. In addition, physical properties have been improved for easier transport and disposal at landfills. Specific process examples are ProFix™ and Cement Slag, detailed below.

ProFix™. This process was designed for two waste treatment applications:

- where physical sorption is required to take up the excess water in low-solids waste, while producing a hardened product by chemical reaction, a requirement under the 1985 Land Disposal Restrictions (US EPA 1986b); and

- where slightly soluble metal compounds are present in the waste, and could continue to dissolve over time or diffuse out from porous particles. In this case, the advantage is that the slow, continuous generation of soluble silicate provides a reserve capacity that can re-speciate the dissolving metal as "silicates."

The ProFix™ process, because of its high sorptive capacity, porous structure, and high surface area, can also be applied to immobilize organics. Some ash also has sufficient carbon content — up to 5% — to potentially act in a fashion similar to activated carbon.

Cement-Slag. These processes are especially applicable for remediating primary metal refining wastes when practiced at the slag producer's site, where the slag is available at little or no cost, or even with a credit for waste disposal. One form of this process developed by Oak Ridge National Laboratory (ORNL) was designed to immobilize technetium and nitrates. The ORNL demonstrated processes would be applicable to both Tc^{+7} and Cr^{+6} treatment, based on its ability to reduce the oxidation number.

2.1.2.3 In Situ Stabilization and Stabilization/Solidification

Auger-type S/S systems are especially useful in two scenarios:

- where VOCs and odors must be controlled; and

- where the waste to be treated is quite deep, particularly more than 7.6 m (25 ft), the practical limit of backhoe-based systems. They also allow better spatial control of mixing since both horizontal (surface) location and vertical position are known precisely at all times. Reagent introduction and mixing are more uniform than with the more subjective, operator-dependent control in backhoe systems. Finally, auger-type systems are more efficient in sequential treatment operations, such as VOC removal followed by S/S.

2.1.3 Treatment Trains

Stabilization or S/S is frequently the only *treatment* process used in a particular remedial project. Other site activities in the remediation project include site preparation, monitoring, final grading, and cover. In addition, the site might require cutoff walls, liners, leachate collection systems, and other means to limit incursion of rain, surface and groundwater into the site and the egress of leachate from the site. In the case of ex-situ stabilization or S/S, excavation and replacement of treated waste is normally required, and a pretreatment step in the form of waste particle size reduction might be necessary. With in situ stabilization, these latter activities are usually neither required nor feasible, although sometimes in situ stabilization or S/S is followed by excavation and removal to another landfill location.

Other than the general site operations described above, S/S products rarely undergo posttreatment of any sort. However, stabilization is sometimes used as a pretreatment step for subsequent solidification of the waste, if the original waste is not a solid or if the final waste form must have specific physical properties.

2.2 Polymer Stabilization/Solidification

2.2.1 Scientific Principles

Polymer S/S technologies process waste at relatively low temperatures by combining or surrounding wastes with liquid polymers. Cooling or curing of the polymer then produces a solidified final waste form product. Although polymer processes are primarily used to solidify waste, when combined with additives, they can be considered S/S technologies. These technologies are grouped together based on a fundamental similarity in the molecular structure of polymers, which are made of large molecules formed from the union of simple molecules (monomers).

In many respects, however, the stabilization/solidification technologies in this category are more divergent than similar. For example, polymer encapsulation or solidification encompasses a wide range of potential binder materials and several treatment concepts. The binders can be either organic (e.g., polyethylene, vinyl ester-styrene) or inorganic (e.g., sulfur polymer). They can be either thermoplastic or thermosetting. Depending on the nature of the waste, polymers can be applied for either microencapsulation or macroencapsulation. Stabilization/solidification can be accomplished ex-situ or in situ.

Polymers can be grouped in two distinct categories, i.e., thermoplastic and thermosetting, based on the means required for processing:

> *Thermoplastic binders* are materials with a linear molecular structure that repeatedly melt to a flowable state when heated and then harden to a solid when cooled. Polyethylene, sulfur polymer, and bitumen are thermoplastic materials used for waste treatment. Since bitumen is a commercially-available, conventional technology and was covered in detail in the companion monograph (Colombo et al. 1994) it is not included in this volume. Thermoplastics, when used for microencapsulation, are melted, mixed together with waste, and allowed to cool into a monolithic solid waste form in which small waste particles are interspersed within the polymer matrix. For macroencapsulation, the molten thermoplastic is poured into a waste container in which large pieces of waste material have been suspended or

supported. Upon cooling, the thermoplastic forms a solid polymer layer surrounding the waste.

Thermosetting binders are materials that require the combination of several liquid ingredients (e.g., monomer, catalyst, promoter) to polymerize and harden to a solid and which cannot be reversed to a flowable state without destroying the original characteristics. Vinyl ester-styrene, polyester-styrene, and epoxies are examples of thermosetting resins that have been used for waste treatment. Thermosetting resins can be used for microencapsulation and macroencapsulation of waste, as well as for in situ S/S. Since thermosetting polymer processes have been commercially available for more than 20 years, only novel applications of thermosetting polymer technology, such as in situ stabilization, are covered herein.

2.2.1.1 Polyethylene

Polyethylene is an inert crystalline-amorphous thermoplastic material with a relatively low melting temperature. It is produced through polymerization of ethylene gas and the structure of the plastic can be varied to create diverse products with different properties. For example, high-density polyethylene (HDPE) is formed of long polymer chains with relatively little branching, allowing the polymer layers to be closely packed. Typical HDPE densities range between 0.941 and 0.959 g/cm^3 (58.7 and 59.9 lb/ft^3). Low-density polyethylene (LDPE) is produced by inducing a higher degree of chain branching, which keeps the layers further apart. The branches in LDPE occur at a frequency of 10 to 20 per 1,000 carbon atoms, creating a relatively open structure. Typical LDPE densities range between 0.910 and 0.925 g/cm^3 (56.8 and 57.7 lb/ft^3). Low density polyethylene has a lower melt temperature (120°C [248°F]) and melt viscosity than high-density polyethylene (180°C [356°F]), and is easier to process for waste encapsulation applications. Polymer melt viscosity is inversely proportional to molecular weight and is characterized in terms of the melt index, which describes the flow of molten polymer under standard conditions specified by the American Society for Testing and Materials (ASTM 1990). Low-density polyethylene is commercially available with melt indices ranging from 1 to 55 g/10 min.

While polyethylene is a relatively new engineering material, extensive testing to establish long-term durability of LDPE for use in encapsulating waste has been conducted (Kalb, Heiser, and Colombo 1993). Since there are no proven test methods to evaluate long-term durability of any material, the approach taken was to examine potential degradation mechanisms using accepted (e.g., ASTM), short-term tests. Polyethylene resists a wide range of chemicals and solvents, thermal cycling, saturated conditions, and microbial attack. If polyethylene is exposed to radiation doses up to 10^8 rad, cross-linking of polymer chains increases and imparts greater mechanical strength and lower leachability.

2.2.1.2 Sulfur Polymer Cement

Sulfur polymer cement (SPC), also known as sulfur polymer, was developed by the U.S. Bureau of Mines (USBM) in an attempt to create new, commercially-viable construction applications for sulfur produced during the refining of petroleum and in the cleanup of SO_2 stack gases. It was initially used to treat radioactive, hazardous, and mixed wastes by Brookhaven National Laboratory (Kalb and Colombo 1985a) and has subsequently been investigated by the Commission of the European Communities (Van Dalen and Rijpkema 1989), Idaho National Engineering Laboratory (INEL)(Darnell 1991), and Oak Ridge National Laboratory (ORNL)(Mattus and Mattus 1994). It is produced by combining elemental sulfur with readily available and relatively inexpensive chemical modifiers which significantly improve product durability. Elemental sulfur is reacted with 5% (by weight) dicyclopentadiene, which suppresses the solid phase transition in the unmodified material responsible for lowering density and creating an unstable solid. Sulfur polymer cement is manufactured commercially under license from the USBM, and is marketed under the trade name Chement 2000 (Martin Resources, Odessa, Texas).

Despite its name, SPC is a thermoplastic material, not a hydraulic cement. It has a relatively low melting point of 120°C (248°F) and melt viscosity of about 25 centipoise (0.0168 lb/ft-sec), and thus can be processed easily with a simple, heated, stirred mixer. Compared with hydraulic Portland cements, sulfur polymer cement has a number of advantages. The compressive and tensile strengths of SPC can be twice those of comparable Portland concretes. Full strength is reached in a matter of hours rather than several weeks. Concretes prepared using SPC are extremely resistant to most acids and salts. Sulfates, for example, which are known to attack some hydraulic cements, have little or no effect on the integrity of sulfur polymer cement. Sulfur polymer waste forms exposed to gamma radiation doses up to 10^8 rad, did not reveal any statistically significant changes in mechanical integrity (Kalb et al. 1991; Van Dalen and Rijpkema 1989).

2.2.1.3 Thermosetting Polymers

Thermosetting polymers are formed by the polymerization of an unsaturated monomer (e.g., methacrylates), typically by a chain reaction. The reaction is initiated by a chemical catalyst, such as benzoyl peroxide, which is decomposed by thermal energy or the action of a chemical promoter, such as dimethyl toluidine. The decomposition breaks O-O bonds, forming free radicals which have unpaired, highly reactive electrons. The free radicals, in turn, break the double bonds of the monomer and add to it. This process is exothermic and continues rapidly, as more and more monomers add to the chain. The reaction finally terminates when the monomer is consumed or chains meet end-to-end. It can be controlled by temperature (increased temperature accelerates the chain reaction), promoter-catalyst combinations and concentrations, and the presence of admixtures (or waste materials) that can retard or accelerate the set. The gel time is defined as the period during which the resin viscosity increases rapidly and finally can no longer be poured or worked. The gel time can be varied by the manufacturer of the resin or catalyst-promoter.

Thermosetting resins have undergone extensive durability and performance testing for both ex-situ and in situ waste treatment applications. Since they are typically low in viscosity (3 to 300 centipoise [$2 \cdot 10^{-3}$ to $2.02 \cdot 10^{-1}$ lb/ft-sec]), they are readily adaptable to in situ injection in soil (Heiser, Colombo, and Clinton 1992). When combined with waste aggregates to form a polymer concrete, they have excellent mechanical strength (48.3 MPa [7,000 psi] or greater, depending on the type of soil aggregate). In situ polymer concretes formed from contaminated soils are highly resistant to aggressive chemicals (acidic and alkaline environments), thermal and wet-dry cycling, microbial degradation, and radiation doses to 10^8 rad. Low hydraulic conductivity ($<2 \cdot 10^{-11}$ cm/sec [$<7.9 \cdot 10^{-12}$ in./sec]) and leachability have been demonstrated (Heiser and Milian 1994). A disadvantage of thermosetting polymers is that, unlike thermoplastic polymers, once polymerized they cannot be re-worked.

2.2.2 Potential Applications

Broad application to diverse waste streams is one of the primary advantages of polymer S/S processes. Polymer S/S technologies can be used in place of conventional cement grout S/S for production of final waste forms with improved durability and leaching performance. High waste loadings can result in fewer waste forms for storage, transportation, and disposal, providing savings in life-cycle costs compared with conventional technologies (Kalb and Colombo 1985b; Kalb, Heiser, and Colombo 1991a). These technologies can be applied for waste management operations to treat aqueous concentrates, sludges, incinerator ash, ion exchange resins, secondary wastes from offgas treatment, and failed conventional concrete waste forms, as well as environmental remediation applications, including direct treatment of soils, sludges, and debris, and indirect treatment of soil washing and other volume-reduction process residuals.

Application and selection of any S/S technology is based on consideration of a number of factors, such as waste characteristics, waste volumes, treatment and disposal costs, and regulatory requirements. For polymer encapsulation technologies, specific issues that impact selection of the type of polymer (organic vs. inorganic, thermoplastic vs. thermosetting) and the method of treatment (microencapsulation vs. macroencapsulation, in situ vs. ex-situ) are dicussed in this section. These include chemical and physical properties of the waste, ultimate disposition of the treated waste, disposal site waste acceptance criteria, final waste form performance criteria, capital and operating costs, availability of materials, ease of processing, and reliability. These issues are described below.

- *Chemical Properties of the Waste.* Thermosetting polymers require a chemical polymerization reaction to form a solid product; interactions with constituents found in the waste can impede formation of free radicals and adversely impact the solidification process. For example, wastes that contain reducing agents (e.g., reduced metals such as iron), complexing agents (e.g., EDTA) or sorbents (e.g., carbon filter media) can interfere with the effectiveness of the catalyst to initiate and complete polymerization. Thermoplastic polymers do not rely on chemical reactions to form a solid final waste form product, and solidification is assured on cooling. However, thermoplastic polymers cool to a solid when exposed to a large thermal mass (e.g., soil) and therefore are less suitable for in situ applications where thorough penetration is required.

- *Physical Properties of the Waste.* Particle size and distribution, density, and moisture content of the waste can impact polymer S/S processes. In general, wastes that consist of small particles are treated more effectively by microencapsulation. The smaller the waste particles, the greater the ratio of polymer/waste surface area, resulting in improved leachability characteristics. However, microencapsulation of extremely fine particles (e.g., <50 μm), especially those with densities <1.0 g/cm^3, may be limited for some viscous thermoplastic binders. Likewise, particles >3 mm (0.118 in.) are more effectively processed following size reduction. Large particles that are considered debris (>60 mm [2.4 in.]) are more effectively treated by macroencapsulation. Thermoplastic processes operate at temperatures in the range of 120 to 180°C (248 to 356°F), so that moisture contained in the waste is volatilized. In most cases, it is advantageous to pre- treat the waste to remove residual moisture. However, small quantities of moisture (e.g., <2% by weight) can be removed during processing. Although most hazardous metals and radioactive contaminants are not volatile at thermoplastic processing temperatures, some highly volatile species (e.g., mercury) might need to be captured in the offgas or auxiliary treatment. For wastes containing significant concentrations of VOCs (>5% by weight), removal and destruction of the organics are recommended prior to treatment by polymer encapsulation. Some thermosetting resins can tolerate significant levels of moisture by forming a high-shear emulsion of the waste within the polymer prior to solidification. Since the process temperature is maintained below 100°C, small droplets of the moisture are trapped within the final waste form.

- *Requirements for Disposal, Waste Acceptance, and Final Waste Form Performance.* Final waste form performance requirements are dictated by the properties of the waste itself (e.g., radioactive, hazardous, mixed), levels of contaminants, where the waste was generated (government or commercial site), and where the treated waste will ultimately be disposed. Waste form performance criteria issued by the Nuclear Regulatory Commission,

U.S. Environmental Protection Agency (US EPA), U.S. Department of Energy (DOE), and the States, as well as waste acceptance criteria issued by individual disposal sites must be considered. Waste form performance issues that can influence selection of polymer encapsulation technologies, in particular, include mechanical integrity, durability, leachability, biostability, and radiation stability.

- *Cost.* The cost of polymer materials varies widely, from a low of about \$0.26/kg (\$0.12/lb) for sulfur polymer to a high of more than \$14.33/kg (\$6.50/lb) for polysiloxanes and some epoxies. Cost comparisons are further complicated by variations in waste loading efficiencies which affect the number of drums processed for disposal, shipping costs, disposal costs, and processing costs. Therefore, it is particularly important to evaluate fully the cost-effectiveness of specific polymer encapsulation technologies under consideration.

- *Availability.* Most polymers that have been considered or used for waste encapsulation are commercially available. Some polymers, however, are not widely used for other applications and/or are produced by a limited number of suppliers (e.g., sulfur polymer). At any given time, these materials might be in short supply, especially without advance notification to the manufacturer. The demand for some polymers used for waste encapsulation could exceed the current rates of production for conventional applications, and could create viable new markets. However, polymer manufacturers should easily be able to meet any increased market demands.

- *Simplicity of Process.* Polymer processes vary in complexity from relatively simple thermoplastic materials requiring heating/ mechanical batch mixing (e.g., sulfur polymer) to more complicated thermosetting polymers that require precise addition of catalyst and promoters to initiate chemical polymerization. Polymer processing is done at ambient temperatures (thermosetting polymers) or at slightly elevated temperatures (150°C [302°F] for LDPE), reducing or eliminating the need for complex offgas collection and secondary waste treatment, an advantage compared with higher temperature vitrification processes.

- *Reliability.* Although polymers are relatively new engineering materials, polymer processes in general, and thermoplastic processes in particular, have proven reliable in other applications in over 50 years of development. Thermoplastics are routinely used for packaging, piping, mechanical parts, etc., providing a large base of experience and high level of process reliability.

- *Re-Work Capability.* Thermoplastic polymers possess a unique advantage in that they can be readily reprocessed, if necessary, by simply remelting and reforming the waste form. In a similar fashion, processing can be restarted following unplanned shutdowns, by simply reheating materials to a molten state. This unique property of thermoplastic resins also allows the use of recycled plastics from either industrial or post-consumer sources.

2.2.3 Treatment Trains

Thermoplastic polymer S/S processing is achieved by heating the polymer binder to the melting temperature, adding the waste material, and mixing to a homogeneous condition. Depending on the type of polymer, the treatment train may consist of simple, heated batch mixers (e.g., double planetary mixers), extruders (single-screw or twin-screw), and thermokinetic mixers. Thermosetting polymers are processed by adding a small quantity of catalyst and promoter to the thermosetting monomer, adding the waste material, mixing to form a homogeneous blend and allowing time for the polymerization reaction to occur. For in situ applications, low viscosity thermosetting monomers are used. These materials are applied by either flooding the waste (e.g., soil) by a technique known as permeation grouting or by injecting the monomer into the waste under pressure by a technique known as jet grouting.

Particle size, density, and moisture content of the waste limit the use of some polymer technologies. These limitations can usually be ameliorated, however, by implementing appropriate pretreatment technologies, such as drying, emulsifying, size reduction, or agglomeration. In other cases, the polymer processing equipment itself can be modified to improve processibility and final waste form performance.

2.3 Vitrification

2.3.1 Scientific Principles

Vitrification is the class of stabilization and solidification technologies that expose the waste stream to high temperatures (i.e., >1,000°C [1,830°F]) to achieve the treatment objective. In doing so, organic contaminants and combustible materials are pyrolized or combusted. Combustion can occur upon first entering the melter, within the feed pile, in the plenum space, or in the afterburner. Inorganic metals and oxides are converted into a glass, crystalline, and/or slag product. In an oxidizing environment, the waste components are converted to oxides, react together, and form the vitreous product. Decomposition products evolving from the process include water and oxides of carbon, nitrogen, and sulfur, if present in the waste stream. In addition, particulates and semivolatiles might also be generated. The behavior and fate of a majority of the semivolatile metals, such as lead, nickel, antimony, etc., have been compiled by the US EPA (1992a). The processing conditions and final product vary greatly depending on the wastes treated and the final product property requirements.

What generally differentiates the vitrification technologies is the method by which the thermal energy is provided. The primary heating methods used include plasma heating, direct electric heating, fossil fuel combustion, induction, and microwave heating. Technologies based on these heating methods are discussed here, with the exception of plasma heating which has not matured to the point that sufficient information is available to include in this monograph. However, plasma based technologies, including DC and AC graphite arc and traditional plasma torch, are expected to emerge in the next few years. Similar to aqueous S/S and polymer S/S, vitrification can be applied either ex-situ or in situ.

2.3.1.1 Ex-Situ Melters

Electric Melters. Electric or joule-heated melters produce a glass or vitreous product. The composition and cooling history of the product upon discharge determine whether the product will have a vitreous or crystalline nature. Crystalline phases do not usually degrade the leach resistance of the waste form. As described by Colombo et al. (1994), the glass components fall into three categories: glass formers, stabilizers, and fluxes. Glass former oxides (principally oxides of silicon, aluminum, phosphate, and lead) form the skeleton or glass network. Stabilizers, such as transition metal oxides and alkaline and rare earth oxides, affect the durability, electrical conductivity, and viscosity properties of the glass. The fluxes, principally alkali metal oxides have a very strong affect on the durability, electrical conductivity, and viscosity properties of the glass. Between room temperature and approximately 500°C (932°F), glass is not electrically conductive. However, above 500°C (932°F), the glass structure begins to "relax," and the ionic species, i.e., the alkali metals, become mobile. In the presence of an alternating electric current field, the ionic species will generate heat according to Joule's Law, $P = I^2R$.

Combustion Melters. Combustion melters can produce a vitreous, crystalline, or slag product depending on the design of the melter. Combustion melters for stabilization and solidification of waste streams are either based on commercial glass melter designs, kiln furnace designs, or adaptation of systems originally designed as furnaces for heat generation. The latter are emphasized in this monograph because of their emerging technology status. Combustion melters burn a fossil fuel such as natural gas, pulverized coal, or fuel oil, over the top of the waste and product materials. Thermal energy is transferred primarily through radiant heat transfer. Waste glass is typically very dark which results in all radiant heat energy being adsorbed within the first few millimeters of the molten glass surface. Therefore, to be efficient, it is necessary to either maintain a relatively thin waste and glass product layer or actively mix the process to facilitate natural conductive and convective heat transfer. Because joule heating is not used, the electrical conductivity of the glass is not important. As a result, combustion melters allow somewhat more freedom in defining the glass composition, although the alkali oxide content strongly influences the viscosity of the molten glass and resistance to leaching, i.e., chemical durability.

Induction Melters. Induction melters are similar to electric melters in that heating is achieved by joule-heating the glass, although the methods differ. In an electric melter, the electric potential is applied across the glass. In induction melting, a magnetic field potential is applied across the glass. The variation in magnetic field causes a change in magnetic flux passing through the glass according to Lenz's Law. Lenz's Law, restated to apply to this case, states that when an electromotive force is induced within molten glass by any change in the relation between the molten glass and the magnetic field, the direction of the electromotive force produces a current which has a magnetic field that opposes the change. These eddy or induced currents are converted to heat through the joule-heating effect. A thorough discussion of the application of induction heating for industrial uses is given by Orfeuil (1987). As is true for electric melters, heat transfer occurs primarily through natural convection and conduction from the molten pool to the waste material unless active mixing of some means is employed.

Microwave Melters. In microwave melting, a container, conveyor, or furnace holding the waste is connected to a microwave generator via a wave guide. The microwaves are transported down the guide and enter the processing chamber where the microwaves couple to the waste and product. The melting process is achieved through three primary mechanisms:

- frictional heat caused by the vigorous vibration of dipolar molecules due to oscillation of the electromagnetic field;

- frictional heat caused by the vigorous vibration of magnetic materials due to oscillation of the magnetic component of the field; and

- generation of heat by electrically-conductive materials due to the current generated by the electrical component of the field.

Standard systems sold within the U.S. use 915 or 2,450 MHz microwave energy. The penetration, or effective heating depth, of the microwaves varies depending on the material being treated. Generally, microwaves will penetrate 5 to 10 cm (1.97 to 3.94 in.) through materials having a high water content and on the order of tens of centimeters for other materials (Orfeuil 1987).

2.3.1.2 In Situ Vitrification

Both in situ and ex-situ electricity-based vitrification share the same fundamental scientific basis; i.e., electricity is passed through molten silica-based media at levels that generate sufficient heat (joule heating) to melt adjacent media. The resulting glass and crystalline product incorporates nonvolatile and noncombustible waste species in a highly durable waste form. For in situ vitrification, volatile and combustible organic waste species are pyrolyzed and/or vaporized below grade. Pyrolysis products oxidize at the melt surface. The gaseous products of these reactions are directed to an offgas treatment system for additional polishing, removal, or destructive treatment to meet air emission requirements. An ex situ glass melter allows greater flexibility in controlling additives to the melt to control primary parameters, including electrical conductivity, melting temperature, viscosity of the hot glass, and quality of the resulting glass product than is possible with in situ vitrification.

In situ vitrification (ISV) has several important advantages compared to ex-situ vitrification processes, including the following:

- ISV can operate at higher temperatures (1,600° to 2,000°C [2,900° to 3,600°F]) than melters (1,100° to 1,400°C [2,000° to 2,550°F]). To provide for an acceptable life expectancy for an ex-situ melter's refractory lining, chemical fluxes must be added to the media processed in melters so as to lower the melt temperature. Even then, melters must be periodically shut down and relined. ISV is free of this limitation since it does not use a refractory lining. This normally allows ISV to be applied to soil and other earthen media without the addition of costly fluxes.

- The higher operating temperature of ISV processing produces a higher quality product compared to the typical ex-situ melter product, in terms of chemical leach resistance and weatherability.

- ISV processing results in greater net volume reduction since additives generally are not necessary.

- No corrosion of refractory bricks occurs due to contact of the bricks with unoxidized metals because ISV uses the surrounding soil for containment of the melted waste-soil mixture instead of bricks.

- ISV is capable of treating most materials without pretreatment, and without the need for size reduction to allow feeding to the melt.

- Because of the greater depth at which elemental metals are incorporated into the melt (up to 20 feet), ISV retains a higher fraction of metals than melters and other vitrification processes which use much shallower bed depths and typically treat contaminants at the surface of the melt. Melting at the surface results in higher losses of metals to the offgas (e.g., plasma melting loses most of the metals).

- ISV processing produces a monolithic waste form that has less surface area for weathering or chemical leaching exposure than melter products.

- The on-site and in situ nature of ISV processing improves occupational, public, and environmental safety compared to melter processing, which requires significantly more handling and transport of contaminated media.

The ISV technology can be applied in four basic configurations, including:

- in situ — the contaminated materials are treated where they presently exist in the ground;

- staged in situ — contaminated materials are partially or completely consolidated or relocated to treatment cells for treatment above, below, or partially below grade;

- stationary batch — materials are melted in one location, the vitrified product is removed after treatment, and the cycle is repeated over and over; and

- stationary continuous — wherein processing is performed in one location with materials being continuously fed to the melting zone and treated molten material being continuously removed.

It should be noted that the first two configurations listed above involve leaving the melts in place and moving the equipment between melts to treat large areas. The latter two configurations involve moving the materials to be treated and removing the vitrified product while leaving the equipment in a stationary processing location.

In situ vitrification melts media that are contaminated with, or are in close proximity to, waste materials that must be destroyed or immobilized. The media and/or waste must have sufficient alkali content (1.4 to ~15%) to ensure the proper balance between electrical conductivity and melting temperature. Too much alkali (>15%) increases the electrical conductivity to the point that insufficient heat would be developed when operating at the electric current limit of the equipment. Too little alkali can result in undesirable, high melt temperatures or insufficient electrical conductivity. Most natural, earthen materials contain sufficient quantities of alkali to allow efficient ISV processing. The rare cases of low alkali content or insufficient soil can be overcome by injecting an alkali-bearing solution into the soil or by simply mixing additional soil or chemicals into the media.

2.3.2 Potential Applications

Vitrification technologies are attractive because they process a variety of waste types and combinations of wastes. In addition, nearly every element in the periodic table is soluble to some extent in glass. Table 2.1 identifies the relative solubilities of many of the elements in silica-based glasses (US EPA 1992a). By choosing an appropriate glass system — silica, alumina, phosphate, or lead-based glasses — waste loadings can be maximized, thereby reducing the final product volume that requires disposal. Because vitrification technologies can simultaneously accommodate a large number of elements, it also provides the flexibility for treating multiple or poorly characterized waste streams. Depending on the vitrification technology, metals, combustibles, organics, and inorganics can be processed to varying degrees. In general, waste streams that are essentially organic liquid waste are more efficiently treated using standard incinerator technology; rather than vitrification technologies. Also, of the listed metals, mercury cannot be processed by any of the vitrification technologies due to its relatively low vaporization temperature. Mercury will be essentially completely volatilized from the waste material and discharged to the offgas treatment train. Wastes containing the remaining listed metals can be treated by one or more of the vitrification technologies with varying degrees of efficiency.

Table 2.1
Approximate Solubility of Elements in Silicate Glasses

Less than 0.1% (by weight)	Ag, Ar, Au, Br, H, He, Hg, I, Kr, N, Ne, Pd, Pt, Rh, Rn, Ru, Xe
Between 1 and 3% (by weight)	As, C, Cl, Cr, S, Sb, Se, Sn, Tc, Te
Between 3 and 5% (by weight)	Bi, Co, Cu, Mn, Mo, Ni, Ti
Between 5 and 15% (by weight)	Ce, F, Gd, La, Nd, Pr, Th, B, Ge
Between 15 and 25% (by weight)	Al, B, Ba, Ca, Cs, Fe, Fr, K, Li, Mg, Na, Ra, Rb, Sr, U, Zn
Greater than 25% (by weight)	Si, P, Pb

Adapted from US EPA 1992a

2.3.2.1 Ex-Situ Melters

Electric Melters. Soils containing organics, pesticides, combustibles, heavy metals, and radioactive contaminants are all amenable to treatment by electric melter technology. As soils differ in alkali content and many contaminated soil areas also contain past evaporation and retention pond materials that are high in clays, it is necessary to obtain representative samples for analysis to determine the relative concentrations of the primary glass-making compounds. Large rocks must be crushed or removed from the process stream if uncontaminated. Refractory organic contaminants might require secondary thermal combustion units in the air pollution control train to achieve the destruction and removal efficiency required to meet the facility discharge requirements.

Electric melters have been proposed for treating buried waste; however, the waste would have to be more thoroughly characterized and sorted than would be required for other vitrification processes. Buried wastes consisting predominantly of combustible materials, such as paper and plastics, can be processed. However, the process would not be as efficient as combustion or plasma vitrification systems. Buried wastes that are high in metals would require removal of the metals. Because of the rate of oxidation of metals in the glass, only trace, finely-divided metals content should be considered. Methods have been proposed to either oxidize the metal during processing to bring it into the glass or allow a metal phase to collect in the melter that could be drained from the melter as a separate product. However, neither of these have been demonstrated. Therefore, electric melters have limited application for this waste type.

Electric melters are applicable to process and industrial waste streams. Electric melters can process wastes in dry, slurry, or liquid form. Generally, waste streams high in inorganic salts and oxides would be most amenable to treatment by electric melters. The volume of glass product depends on whether the targeted waste stream contains a large percentage of the glass formers, stabilizers, and fluxes required to achieve the final glass composition. In addition, pretreatment of the waste stream might be required to make it competitive with applicable nonthermal treatment options. For example, dilute streams should be concentrated as much as possible to minimize the water load. This will make the process more efficient and also reduce the fraction of entrainment into the offgas treatment train that would be generated by active boiling.

The efficiency of the vitrification is unaffected by the isotopic nature of radioactive waste constituents. Within the U.S., radioactive wastes include contaminated soils, buried wastes, process wastes streams, and wastes resulting from building and equipment demolition and decontamination. The electric melter was developed specifically to treat DOE defense high-level wastes (Chapman and McElroy 1989). Radioactive wastes will, in general, have the same compatibility for electric melter processing as the waste types described above. There are some radioactive isotopes — tritium, carbon-14, and isotopes of iodine — that will not be immobilized due to their gaseous state at processing temperatures. If the concentration of these isotopes exceeds stack release limits, they must be removed prior to vitrification or captured by offgas treatment equipment and disposed as a secondary waste.

Combustion Melters. Combustion melters are similar to electric melters in their applicability to contaminated soil. However, combustion melters expose the waste to higher instantaneous temperatures, leading to a lower retention of many alkali metals and heavy metals, such as lead, zinc, and cadmium. However, depending on the metals concentration, ultimate retention should be possible by recycling the metals back from the offgas treatment train.

Combustion melters are similar to electric melters in their applicability to buried waste with some clarifications. Combustion melters should be more applicable to buried wastes containing predominantly combustible materials. However, combustion melters place maximum size restrictions on the feed stream materials. Depending on the process, it might be necessary to crush, shred, pulverize, or slurry the materials for delivery to the melter. Large items, such as drums, equipment components, etc., would be excluded. Combustion melters typically do not have a significant glass accumulation capacity; resulting in a short residence time within the melter. Therefore, they would be preferred over other ex-situ vitrification processes if the waste has a high concentration of ferrous and nonferrous metals. These metals would be incorporated in the glass as metal inclusions and should have little effect on the final glass properties.

Combustion melters are similar to electric melters in their applicability to process and industrial waste streams with some clarifications. Predominantly organic liquid wastes are compatible with combustion melters. Combustion melters expose the waste to higher instantaneous temperatures leading to a lower retention of many alkali metals and heavy metals, such as lead, zinc, and cadmium. Ultimate retention should be possible by recycling the metals from the offgas treatment train.

Combustion melters are similar to electric melters in their applicability to radioactive waste. Process wastes that are primarily salt solutions are more difficult to process because of excess volatility of the alkali metals and heavy metals. Ultimate retention should be possible by recycling the metals back from the offgas treatment train.

Other Melters. Induction and microwave melters are similar to electric melters in their applicability to contaminated soils, buried waste, process and industrial waste, and radioactive waste. A benefit of microwave melters is that they can process a wide range of waste compositions, since the electric conductivity of the waste is not important. However, because microwave melters do not have a large glass holding capacity, the incoming waste must be more homogeneous to assure an acceptable glass product. Induction melters can operate at temperatures as high as 2,800°C (5,070°F); making them very applicable to waste streams primarily consisting of metallic waste (i.e., buried waste).

2.3.2.2 In Situ Vitrification

Contaminated Earthen Media (Soil, Sediment, Tailings). The ISV technology applies to contaminated earthen media containing highly variable amounts of sand, silt, and clay. Even rocky soils can be melted by the process (Thompson, Bates, and Hansen 1992).

The ISV process is tolerant of multiple small voids in the soil of up to 0.07 m³ (2.5 ft³) each. Until further research resolves questions regarding the effects of voids, larger voids should be collapsed or filled to preclude the possibility of generating large bubbles in the melt. Large bubbles, when released at the molten surface, can cause excessive agitation of molten glass and release of heat inside the hood.

In situ vitrification is generally applicable to most soils, regardless of moisture content. Soils and sludge ranging from 4 to 70% moisture (by weight) have been successfully vitrified (Thompson, Bates, and Hansen 1992). Geosafe Corporation, the sole U.S. licensee for the ISV technology, states that saturated soils and slurries can be successfully vitrified. The amount of water associated with silty soils or non-swelling clays is important when determining the economic feasibility of ISV in those applications. The technology, however, might not be economically feasible in soils that lie within a permeable aquifer (i.e., with a hydraulic conductivity of greater than 10^{-4} cm/sec), unless combined with a groundwater diversion or pumping technique to limit the rate of water recharge to the treatment zone.

Because most soils and sludges are naturally composed of glass-forming materials, such as silica, they can generally be processed by ISV without modification. A total alkali content (i.e., combined Na_2O, Li_2O, and K_2O content) of 2 to 5% is desirable, however. Alkaline oxides carry the electrical current between electrode pairs when soil is in the molten state. Weathered soils with less than 1.4% (by weight) alkaline oxides require addition and mixing of alkaline materials to lower the melting temperature and raise electrical conductivity (Buelt et al. 1987). Excessive amounts of alkali can pose processing problems by lowering the electrical resistance, which reduces the melt temperature, thereby impacting the quantity of vitrified product obtained.

The ISV process has been demonstrated at depths of up to 5.8 m (19 ft) in relatively homogeneous soils. The achievable depth, however, can be limited under certain heterogeneous conditions, such as the presence of a rock or gravel layer, which inhibits melting, or if a soil layer with a significantly higher melting temperature than the overlying material exists. Melting depths of 4.3 m (14 ft) and 5.2 m (17 ft) have been attained when rock layers existed at those depths. The relative density of the soils to be processed also influences the achievable melt depth. Higher-density soils require more time and energy to be processed due to their higher bulk heat capacities (Thompson, Bates, and Hansen 1992). The vitrification process eliminates the porosity of the soil, thereby increasing its density by 33% to 100% in the vitrified form, according to Geosafe Corporation.

ISV is highly effective at immobilizing heavy metals and other inorganic contaminants in buried waste. The majority (70 to 99.9% by weight) of heavy metals, such as arsenic, lead, cadmium, barium, and chromium, are retained and immobilized in the vitrified product (Thompson, Bates, and Hansen 1992). The retention efficiency of metals depends on:

- vapor pressure;

- solubility in the molten media; and

- depth of melt.

Metals evolved from the melt are collected by the offgas system and either recycled to future melts or disposed separately. Nitrate salts are decomposed by the process, releasing NO_x to the offgas system. Mercury may be removed and collected by the offgas system for reuse or disposal.

The high processing temperature (up to 2,000°C [3,630°F]) of ISV destroys hazardous organic chemicals by pyrolysis. See Table 2.2 for data from the ISV demonstration at the Wasatch Chemical Superfund Site in Salt Lake City, Utah. Organic concentrations of up to approximately 10% by weight in the soil can be processed with the current technology. (The limit is based on the ISV system's ability to accommodate the heat loadings resulting from the exothermic oxidation of the pyrolysis products. The limit also depends on the distribution and heating value of organics in the soil.) The small amount of organic contaminants not destroyed by the process (between 0.01 and 1% by weight) is removed from the soil through diffusion, thermal convection, and the negative pressure induced in the offgas hood and is processed by the offgas treatment system. In situ vitrification should not be applied to unexploded ordnance or highly reactive materials, since little theoretical or experimental work has been directed toward these types of materials. However, ISV has been successfully tested and proven safe and effective on explosives-contaminated soils from the Mead Nebraska Ordnance Plant (Cambell, Schultz, and Cichelli 1994). Empirical data show that VOCs such as carbon tetrachloride and trichloroethylene can be effectively removed by ISV by a combination of pyrolysis, oxidation, and offgas treatment (Shade et al. 1991). These data indicate that about 97% of these VOCs are destroyed, about 3% are captured by the offgas system, and less than 0.1% remain in the soil surrounding the glass melt.

With recent improvements in the electrode-feed system, which permits electrodes to be inserted as the soil is melted downward, high scrap-metal concentrations can be processed by ISV. Vitrification of up to 37% of elemental metals in soils has been demonstrated. The metal melts, and because of its higher density, forms a molten pool at the bottom of the pool of molten soil. The molten metal layer can create electrical short circuits between the electrodes, but by retracting the electrodes a few centimeters above the molten metal, short circuits can be overcome.

High concentrations of concrete, asphalt, rubble, rock, scrap metal, plastics, wood, tires, and other debris (up to 50% by weight) can generally be processed by ISV if all other operational constraints are met (Thompson, Bates, and Hansen 1992). Monolithic debris and other structures that might impede the release of water vapor from beneath the molten soil to the soil's surface should only be vitrified when sufficient analysis of the effects of the debris shows ISV to be safe and efficient.

Table 2.2
Hazardous Organic Chemical Destruction and Removal Effectiveness Using In Situ Vitrification

Pre- and Post-ISV Contaminated Soil and Adjacent Surrounding Soil Levels with Cleanup Criteria

Indicator Chemical	Pretreatment (ppb)	Posttreatment (ppb)	Surrounding Soil (ppb)	Regulatory Limit (ppb)
TCDD Dioxin	11	≤ 0.12 *	≤ 0.0045	20, < 1(TCLP)***
2,4-D	34,793	≤ 20 **	ND	N A
2,4,5-T	1,137	< 14 **	ND	N A
4,4-DDD	52	ND	ND	28,000
4,4-DDE	3,600	ND	≤ 2.4	19,000
4,4-DDT	1,090	ND	ND	19,000
Total Chlordanes	535,000	≤ 80	≤ 83.4	7,000
Heptachlor	137.5	ND	ND	2,000
Hexachlorobenzene	17,000	ND**	ND	7,000
Pentachlorophenol	272,918	≤ 10.3 (non-TCLP)	≤ 1.2	< 10 (TCLP)**
Trichloroethene	36,875	ND	ND	103,000
Tetrachloroethene	< 100	ND	ND	22,000

Stack Emission Performance Data

Emission Values (lb/hr)

Indicator Chemical	Run 1	Run 2
TCDD Dioxin	< 2.46E-9*	< 2.09E-9
2,4-D	< 3.09-E-6	< 1.35E-4
2,4,5-T	< 6.19E-7	< 2.70E-5
4,4-DDD	< 6.19E-8	< 2.09-E-6
4,4-DDE	< 6.19E-8	< 2.09E-6
4,4-DDT	< 6.19E-8	< 2.09E-6
Total Chlordanes	< 6.19E-8	< 1.05E-6
Heptachlor	< 6.19E-8	< 1.05E-6
Hexachlorobenzene	< 6.19E-8	< 1.05E-6
Pentachlorophenol	< 1.24E-6	< 2.70E-6
Tetrachloroethene	N A	< 1.10E-04
HCL	< 0.05	< .0046

*< Values indicate contaminant was not detected at the reported detection limit
**Denotes contaminant was found to be present at low levels in two of the six samples taken
***TCLP analysis performed on the product from the dioxin area only with all results found to be below detection limits

Although field data for vitrifying buried combustible materials are limited, ISV has been used successfully to process more than 80 buried creosote timbers, each measuring 3.6 m (11.8 ft) long by 15 cm (5.9 in.) square, in a single, large-scale setting (Luey et al. 1992). Based on the heat-removal capabilities of existing ISV equipment, combustible inclusions of up to 10% (by weight) can be processed.

The ISV process is not applicable to sealed containers, including empty tanks, 208 L (55 gal) drums, and 3.8 L (1 gal) paint cans containing liquids because of the potential for disruption of the melt due to the release of pressurized gases when these containers are breached. Similar concerns exist for pockets of liquids. For many landfills, corrosion can degrade sealed containers, largely eliminating the potential for transient releases of trapped gas related to source liquids. The potential for melt disturbance due to rapid depressurization of sealed containers of liquids can be alleviated using dynamic disruption and compaction.

ISV is capable of treating underground pipelines, cribs, and drain fields. Tanks (including residual wastes) can be treated in place after preconditioning (e.g., filling with earthen material to remove voids and grouting to prevent rapid conversion of liquids to the gaseous state). Building demolition debris (e.g., concrete and steel), if backfilled with soil, can be directly treated by ISV.

Geosafe, together with its Japanese partner, ISV Japan Limited, is treating industrial wastes in a stationary-batch ISV system. The ISV and joule-heated melter technologies have been adapted by Battelle Pacific Northwest National Laboratory to treat newly-generated process and industrial wastes, including municipal wastes. The adapted technology, called Terra Vit, uses inexpensive refractories and construction techniques. The Terra Vit system, which is constructed below or at the ground surface, greatly reduces the large capital investment associated with standard melter systems. Molten glass and molten metal (if created in the vitrification system) can be separately recovered and cast into engineered shapes for various uses. Terra Vit technology is currently being licensed by Geosafe.

The ISV process has been proven effective for treating soils and buried wastes contaminated with radionuclides, including transuranic materials and fission products. Criticality limits of approximately 30 kg (66 lb) of plutonium per setting have been established for vitrifying soils containing transuranic materials (Thompson, Bates, and Hansen 1992). Thus, soils contaminated with thousands of nCi/g of transuranics can be treated safely with ISV. When high concentrations of ^{137}Cs exist (i.e., when several curies are present at a single setting), special measures must be taken to collect and remove the small percentage (<3% by weight) of cesium that volatilizes to avoid undesirable levels of worker exposure to ionizing radiation.

In situ vitrification is effective for treating mixed wastes. Vitrification processes in general have advantages for such waste because of their ability to simultaneously destroy organics and immobilize inorganics/radionuclides.

2.3.3 Treatment Trains

2.3.3.1 Ex-Situ Melters

Vitrification process equipment provides final and total treatment of a targeted waste stream. The treatment process can be transportable, field erectable, or a fixed facility. A ex-situ vitrification block-flow diagram is provided in Figure 2.3. The waste stream to be treated is delivered to a staging area and is analyzed to determine composition and consistency, if not previously analyzed. This information is used to determine if glass forming additives are needed. The types and amounts of glass forming additives depend on the waste stream — they might not be required at all or they might account for a majority of the final vitrified product.

Pretreatment of the waste occurs next. Large objects are removed to be handled separately or size reduced. Dry feedstocks could then be crushed, shredded, or pulverized as required depending on the melter process. Any required glass forming additives can be added to the feedstream at this time. The glass formers are stored in large bins from which they can be metered to the melter or to a blend tank to be mixed with the waste. If the feedstock is a dilute slurry, dewatering can be performed at this time if determined to be economically favorable. Glass former additives can be blended with the aqueous feedstream or added to the melter as a separate stream. Usually, if the glass formers are added to the liquid stream, less water can be removed prior to blending than would be necessary if the glass formers were added separately. Finally, any recycle streams originating from the offgas treatment equipment are usually blended with the waste rather than being fed to the melter as a separate stream.

Figure 2.3
Ex-Situ Vitrification Block-Flow Diagram

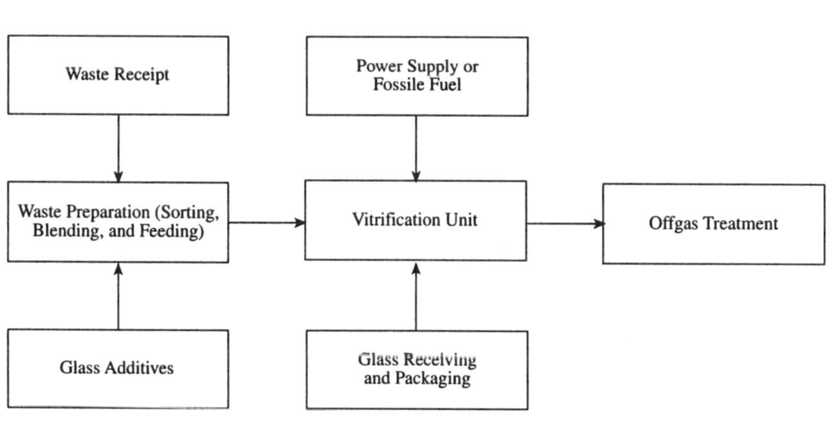

The feed stream can be introduced into a melter in many ways. For electric and microwave melters, the material is dropped directly on top of the glass batch. Therefore, the feed can be augured, pumped, pneumatically transferred, or introduced into the melter in combustible containers. Combustion melters must be fed pneumatically, or the feed is sprayed into the primary melting chamber as a slurry. Little or no evidence supports liquid feeding of induction melters. Therefore, at this time it is assumed that dry feeding is the primary feeding method for these systems.

The size and complexity of the offgas treatment train depends on the waste stream and on the melter type. Combustion melters produce the largest, noncondensable offgas volume because of the combustion gases and excess air normally used to assure complete combustion. The air pollution control industry provides a wide variety of technologies to treat offgases discharged from thermal treatment systems. The waste stream characteristics and treatment requirements will define the best combination of equipment. The reader is encouraged to refer to available air pollution control (APC) publications and literature; as well as contacting APC vendors to identify available options. In general, treatment equipment may be required to treat the following offgas constituents:

- steam;

- acid gases, e.g., HCl, HF;

- halogens;

- SO_x;

- NO_x;

- hydrocarbons;

- particulate "dust";

- gas-phase radionuclides; and

- soluble and insoluble aerosols.

The glass product can be handled in many ways. The product can be cast into containers forming monoliths, air- or water- quenched forming a cullet or type of aggregate, processed through mechanical handling machines to make shapes of uniform size, or cast into molds and heat treated to make products that can be marketed if determined to be nonhazardous.

2.3.3.2 In Situ Vitrification

The ISV treatment system consists of an electric power transformer and cabling, an offgas collection hood, a process control system, and an offgas treatment system (Figure 2.4). The transformer provides two-phase alternating current to the graphite electrodes at the appropriate voltage and current. The offgas hood collects emissions escaping from the treatment zone and provides physical support for the electrodes. The hood is a dome-shaped structure that completely covers the area to be treated. A slight negative pressure is maintained in the offgas hood to contain gases, which are piped to the offgas treatment system. The entire ISV system is monitored from a process control room where electrode power consumption, offgas temperature, hood vacuum, and other system parameters are monitored and controlled.

Several modular unit treatment processes comprise the typical offgas treatment train. The train can include a quencher, scrubber, mist eliminator, heater, filter, activated carbon, and a thermal oxidizer. The choice of treatment units depends on the contaminants and concentrations present at the site. A large-scale offgas treatment system used by Geosafe Corporation was designed to ensure that process air emissions are in compliance with regulatory permit limits. Gases are drawn through the system by an

52 m³/min (1,800 scfm) blower. Automatic valving allows ambient air to enter the hood so that a negative pressure (vacuum) of -1.3 to -5.1 cm (-0.5 to -2.0 in.) of water is maintained. As the rate of gases evolved from the treatment zone changes, the valving controls the amount of ambient air drawn into the system to maintain a relatively constant, negative pressure. Potentially contaminated gases are thus prevented from leaking from the hood.

Figure 2.4
In Situ Vitrification System

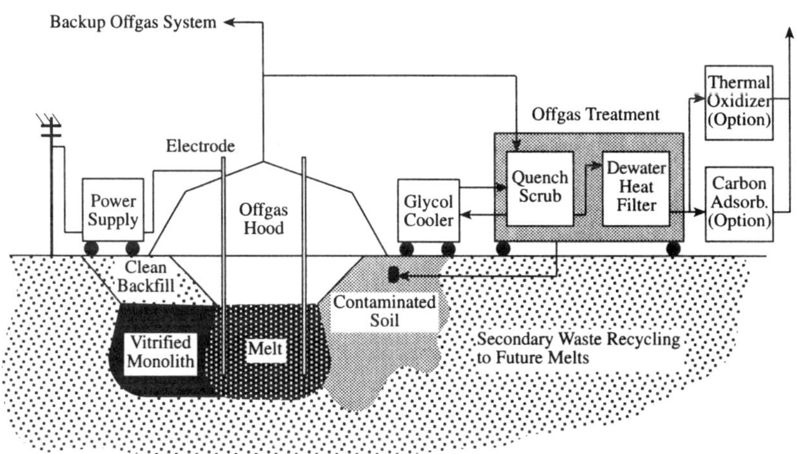

Typically, the volume of gases entering the treatment system consists of greater than 99% ambient air and less than 1% gases evolved from the treatment zone. This ratio varies depending on the soil and contaminants being treated. The dilution effect of ambient air and high combustion efficiency typically results in low levels of hazardous gases within the hood enclosure.

The offgases enter the treatment system at temperatures ranging from about 100 to 400°C (212 to 752°F). The offgases are first quenched to about 80°C (176°F) and then scrubbed in a high efficiency, dual-stage venturi scrubber. A typical scrubber removes 97% of particulate matter with diameters greater than 0.5 μm (0.02 mils).

The quencher and scrubber perform the majority of the offgas cleanup task. Condensable materials condense as mists in the quencher. Most mists and solid particulates are then removed from the offgas by the scrubber. An impingement-type mist eliminator usually follows the scrubber to remove water droplets. The offgas stream is then heated to raise its temperature above the dew point before it passes through a high efficiency particulate air (HEPA) filter and activated carbon filter(s) or thermal oxidizer. Dew-point control prevents "wetting" and blinding the HEPA filter, which removes 99.97% of particulate matter with diameters of 0.3 μm (0.01 mils) or greater.

The gases may finally be passed through activated carbon or a thermal oxidizer to remove or destroy any organics that escaped the quencher and scrubber. The activated carbon or thermal oxidation systems can be designed to remove at least 99.9% of organics in the offgas. Two options exist for treating the scrubber solution:

- return it to the site for retreatment by ISV; or

- provide additional treatment (e.g., solidification) in accordance with disposal requirements.

The method of additional treatment depends on the type and concentrations of contaminants collected in the scrubber solution.

DESIGN DEVELOPMENT

This chapter provides information essential for the engineering design and implementation of Stabilization/Solidification (S/S) treatment technologies. Each of the major categories of S/S technologies are discussed in turn; first, Aqueous S/S, followed by Polymer S/S, and Vitrification with ex-situ and in situ applications of each. For each of these categories and the various modifications, information is provided on the process capability, basis of design, equipment design and selection, process modifications, associated pre- and posttreatment processes, process controls and instrumentation, safety matters, unique factors to be addressed in specification preparation, cost data, design validation, permitting requirements, and performance measures. Each section concludes with a design checklist.

While critical design issues are addressed, this information is intended only as a first step in the engineering implementation process. The reader is encouraged to review the references cited and contact developers and vendors of these innovative technologies. In addition, treatability studies to ensure the applicability of the selected technology for the specific waste stream are recommended and pilot-scale testing is useful in cases where little prior engineering experience exists. For in situ technologies, performance assessment modeling will assist engineers and project managers in evaluating potential behavior of contaminants following treatment.

Technology-specific safety issues are discussed in this chapter. A generic safety issue that should be addressed for any treatment technology is the need for an emergency contingency plan. This plan describes how to recognize emergency or critical situations and who, when, and how facility personnel, local agencies, and the community will be notified during such events. Generally, contingency plans cover routine operation and maintenance inspections, emergency operations, preparedness and prevention requirements, and evacuation procedures. Good communications with local emergency personnel and training are critical to effective implementation of contingency plans.

Although a comprehensive review of regulations governing S/S waste treatment is beyond the scope of this book, key regulatory changes that have been implemented or proposed are reviewed here to provide perspective on technology implementation issues. Federally-mandated cleanup standards under the Resource Conservation and Recovery Act (RCRA), Comprehensive Environmental Response, Compensation, and Liability Act (CERCLA, commonly known as the Superfund Act), and other legislation provide the impetus and technical guidance for waste treatment and environmental restoration.

U.S. regulatory standards for treating hazardous and mixed wastes, however, are continuing to change, requiring waste generators and treatment vendors to continually monitor the regulations and confirm that treatment technologies meet existing and proposed criteria. In some cases, changing regulations require new treatment options be developed, tested, and implemented. A dynamic regulatory climate can thus provide the catalyst for technology development.

The Hazardous and Solid Waste Amendments (HSWA) to the RCRA, largely prohibit the land disposal of untreated hazardous wastes that do not meet treatment standards established by US EPA under section 3004(m). Once a hazardous waste is prohibited, the statute provides only two options for legal land disposal: (1) meet the treatment standard for the waste prior to land disposal, (2) or dispose of the waste in a land disposal facility that has been found to satisfy the statutory no migration rule. A no migration facility is one from which there will be no migration of hazardous constituents for as long as the waste remains hazardous.

US EPA has established treatment standards as specified technologies, as constituent concentration levels in treatment residuals for Listed and Toxicity Characteristic (TC) waste defined in 40 CFR 261.3, or both. When treatment standards are set as levels, the regulated community may use any technology not otherwise prohibited (such as impermissible dilution) to treat the waste. Treatment standards were initially based on the levels achievable by the Best Demonstrated Available Technology (BDAT) for treating the waste. For treatment of hazardous debris defined in 40 CFR 268.45, a series of alternative treatment standards to remove, destroy, or immobilize contamination are required.

US EPA has continued to amend the lists of hazardous wastes and constituents that must be treated and the standards that they must meet for land disposal. On August 18, 1992, a "debris" rule (US EPA 1992b) was issued that established alternative treatment standards for hazardous debris. On May 24, 1993, the "Emergency Rule" was issued that required certain ignitable or corrosive characteristic wastes to be treated to meet more stringent treatment standards for all hazardous constituents reasonably expected in the waste, not just constituents that initially exceeded the 40 CFR 261 levels for characteristic hazardous waste. On September 19, 1994, and April 8, 1996, the Phase II and III Land Disposal Restrictions (LDR) rules (US EPA 1994b; US EPA 1996) broadened these additional requirements by establishing "Universal Treatment Standards" for most characteristic (other than TC metals) wastes that had concentration-based treatment standards. For the applicable wastes, these rules identified more than 200 "Underlying Hazardous Constituents" (i.e., hazardous constituents not present at the point of generation in amounts that exceed the 40 CFR 261 toxicity characteristic levels) that are now subject to LDR.

The Toxicity Characteristic Leaching Procedure (TCLP) is used to identify toxicity characteristic wastes. However, except for TC metals, the Universal Treatment Standards require Total Constituent Analysis (TCA) rather than TCLP to be used to demonstrate the required constituent level for the treated waste.

US EPA has also proposed risk-based levels at which wastes are no longer considered hazardous for purposes of RCRA subtitle C. In the Hazardous Waste Identification Rule (HWIR)(US EPA 1995d), US EPA proposed such risk-based levels for the majority of hazardous constituents found in listed hazardous wastes. Wastes meeting these risk-based levels before or after treatment could be disposed in facilities not subject to RCRA hazardous waste management requirements. US EPA was required to finalize the HWIR exit levels by December 15, 1996. US EPA additionally proposed to allow the exit levels for some constituents to serve as alternative, risk-based LDR treatment standards satisfying the "minimize threat" standard of section 3004(m) of RCRA. Where these risk-based levels are higher (less restrictive) than current treatment standards based upon BDAT, they will effectively supersede current standards.

Cleanup standards for environmental media contaminated with hazardous substances at CERCLA sites are generally based on site-specific risk calculations to meet an acceptable level of risk rather than technology-based performance levels. The risk is dependent on rates of release, routes of exposure, the expected future use of the property or medium, and the specific chemical constituents involved. However, unless countermanded by the

Record of Decision for the site, RCRA LDR treatment standards are also applicable if hazardous substances removed from the CERCLA site are also RCRA hazardous wastes. Hazardous waste treated and left at the CERCLA site may, or may not, also be subjected to RCRA LDR treatment standards, as determined in the Record of Decision, depending on the specific treatment method and its location at the site.

3.1 Aqueous Stabilization/Solidification

3.1.1 Remediation Goals

3.1.1.1 Stabilization

Phosphate Processes. The primary operating advantages of phosphate processes are low waste volume increase, low cost compared to cement-based processes, and ease of application. Their primary environmental advantage is in the reported low leachability of lead compounds over a wide pH range. In the process's basic form, a solution of phosphoric acid or phosphate salt is mixed with the waste by spraying or combining in a mixer. No other physical or chemical operations are required.

Phosphate treatment is best exemplified through commercial application at a number of municipal waste-to-energy incinerators using the WES-PHix® process (O'Hara and Surgi 1988). These installations are totally enclosed, in-line systems that are claimed to reduce leaching of lead and cadmium to well below the TC requirements, thereby eliminating the need for RCRA permits. The system can be used with or without the addition of lime to adjust the pH, depending on ash characteristics. Phosphoric acid or any convenient source of water soluble phosphate can be used, at ratios of about 1 to 8% H_3PO_4 based on weight of the ash. This process has reportedly been used to treat more than 2,700,000 tonne (3,000,000 ton) of ash and other particulate material as of 1993. In this application, the phosphate process is considered conventional; however, application to remedial work is in the developing stages. A remedial project that reported treating 11,800 tonne (13,000 ton) of lead slag (WMX Technologies 1993) is an example of the adaptation of phosphate treatment.

Rubber Particulate. A major research effort at Chemical Waste Management (Conner and Lear 1991) used natural soil spiked with 50 hazardous organic compounds to test the efficacy of various additives for stabilizing organics. Spiking levels ranged from 5 to 2,400 mg/kg, with total organic levels between 0.5 and 1.0% by weight. Twelve different cement-based stabilization formulations were tested. Each formulation contained 15% by weight cement and 8% by weight additive to the weight of the waste. All untreated and treated samples were characterized by TCA and TCLP for all 231 RCRA constituents (US EPA 1994b). Summaries of the TCA results for rubber particulate are given in Tables 3.1-3.3. The results are stated both in absolute terms, with comparison to the US EPA Universal Treatment Standard (UTS), and in percent reduction factors as defined by US EPA. Absolute numbers are important for judging how well a process meets regulatory requirements and numerical treatability objectives. Percent reduction is also provided because it is a much easier number to assimilate than comparing raw data, because the degree of reduction is a method of setting treatability objectives in remedial work, and because it is the best way to document efficiency of the treatment process. Test results adapted from Conner and Smith (1993) are given in Tables 3.1 and 3.2 for volatile and semivolatile organics, respectively. Table 3.3 shows the results of testing a new variation of rubber particulate, KAX-100™, obtained from the vendor (Environmental Technologies Alternatives 1995).

Table 3.1
Immobilization of Organic Constituents Using
Rubber Particulate — Volatile Organics

Compound	USEPA Hazardous Waste Code	Universal Treatment Standard (mg/kg)	TCA Before Treatment (mg/kg)	TCA After Treatment (mg/kg)	% Reduction
Benzene	D018	10.00	200	<10	92
2-Butanone	D035	36.00	640	<50	88
Carbon Disulfide		4.81	640	<200	53
1,2-Dichloroethane	D028	6.00	270	<100	44
Methanol		0.75	1219	444	45

Table 3.2
Immobilization of Organic Constituents Using Rubber Particulate — Semivolatile Organics

Compound	USEPA Hazardous Waste Code	Universal Treatment Standard (mg/kg)	TCA Before Treatment (mg/kg)	TCA After Treatment (mg/kg)	% Reduction
Bis(2-Ethylhexyl)Phthalate		28	150	<0.99	99
Cresol	D023-6	3.2	<9.9	<0.99	85
1,2-Dichlorobenzene		6.0	160	5.64	94
1,4-Dichlorobenzene	D027	6.0	147	6.03	93
2,4-Dinitrotoluene	D030	140	226	<0.99	99
Hexachlorobenzene	D032	10	143	6.96	92
Hexachloroethane	D034	30	114	6.45	91
Lindane (gamma–BHC)	D013	0.066	124	<5.0	94
Methoxychlor	D014	0.18	59.5	<5.0	87
Nitrobenzene	D036	14	166	3.66	96
Pentachlorophenol	D037	7.4	233	0.60	99
Pyridine	D038	16	1900	<0.99	99
2,4,5-Trichlorophenol	D041	7.4	200	<4.8	96
2,4,6-Trichlorophenol	D043	7.4	178	<0.99	99
Phthalic Anhydride (as acid)		28	<30	<3.0	85

KAX-50™ rubber particulate TCA reductions are compared against activated carbon and the EPA UTS in Figure 3.1 for several hazardous organic constituents. Figure 3.2 shows the percent reductions in TCA test results for the two additives on the same constituents; these values have been corrected to account for dilution by the additives and the cement binder, so that they represent the real reduction in mobility. Reductions in TCA ranged up to 99%, indicating the immobilization of low-level organics by stabilization is feasible. While the exact binding mechanisms are not yet known, it is evident that many organic constituents are so strongly held that even the organic solvents used for the extraction step in TCA testing cannot remove them.

Table 3.3

Immobilization of Organic Constituents Using Modified Rubber Particulate, KAX-100™ — Volatile and Semivolatile Organics

Compound	USEPA Hazardous Waste Code	Universal Treatment Standard (mg/kg)	TCA Before Treatment (mg/kg)	TCA After Treatment (mg/kg)	% Reduction
Benzene	D018	10	418	< 30	89
N-Butanol		2.6	3350	< 5.0	100
Carbon Disulfide		4.81	83	< 0.25	100
Chloroform	D022	6.0	431	< 10	97
Cyclohexanone		0.75	536	< 5.0	99
1,2-Dichloroethane	D028	6.0	654	< 0.25	100
Ethyl Acetate		33	258	< 0.25	100
Iso-Butyl Alcohol		170	2100	< 5.0	100
Methylene Chloride		30	62	< 0.25	99
1,1,1-Trichloroethane		6.0	550	< 0.25	100
Trichloroethylene	D040	6.0	881	< 100	83
1,1,2-Trichloro-1,2,2-Trifluoroethane		30	9.1	< 0.25	96
Bis(2-Ethylhexyl)Phthalate		28	150	< 1.0	99
Cresol	D023-6	3.2	<9.9	< 1.0	85
1,2-Dichlorobenzene		6.0	160	< 10	91
1,4-Dichlorobenzene	D027	6.0	147	< 10	90
2,4-Dinitrotoluene	D030	140	226	< 1.0	99
Hexachlorobenzene	D032	10	143	< 10	90
Hexachloroethane	D034	30	114	< 10	87
Lindane (gamma–BHC)	D013	0.066	124	< 5.0	94
Methoxychlor	D014	0.18	59.5	< 5.0	87
Nitrobenzene	D036	14	166	< 5.0	95
Pentachlorophenol	D037	7.4	233	< 1.0	99
Pyridine	D038	16	1900	< 1.0	100
2,4,5-Trichlorophenol	D041	7.4	200	< 5.0	96
2,4,6-Trichlorophenol	D043	7.4	178	< 1.0	99
Phthalic Anhydride (as acid)		28	< 30	< 5.0	75

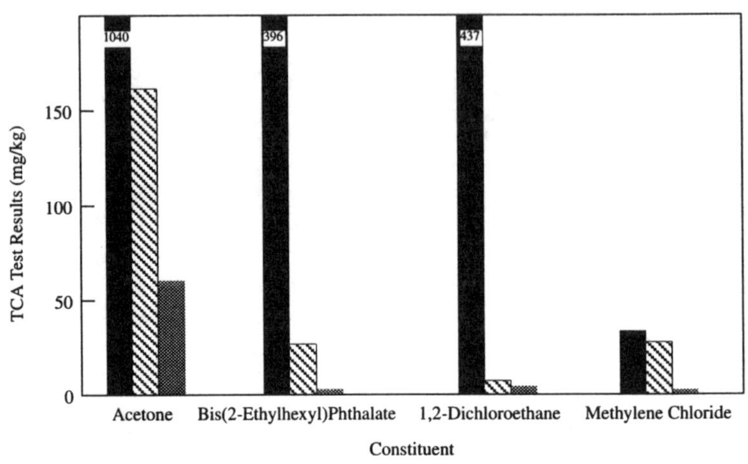

Figure 3.1
Comparison of TCA Results for Carbon vs. Rubber Particulate

Source: Environmental Technologies Alternatives 1994a

However, no "magic bullet" exists in stabilization. Most of the additives described by Conner and Lear (1991) are useful with specific contaminants in specific test methods, and none work best for all. Carbon is effective overall for reduction in TCLP leachability but not for reduction in TCA. Particulate rubber is not as effective in TCLP reduction, but is the only additive that was broadly useful for TCA reduction, especially for the low-volatility compounds. Organo-clays are effective with specific contaminants, and so were other additives tested but not described here.

One surprising result of the above and other recent experimental work (Spence et al. 1990) is that VOCs are not necessarily lost through volatilization during S/S, as was previously thought (Weitzman, Hamel, and Cadmus 1987). Reductions in TCA levels suggest that VOCs sorbed onto or associated with soil particles might be less susceptible than expected to volatilization during stabilization, at least in these slow exothermic reactions, i.e., cement-based systems under relatively static air flow conditions. Some additives, such as rubber particulate, have been shown in independent studies (Environmental Technologies Alternatives 1994b), to substantially reduce the evaporation rate of VOCs so that air pollution is minimized. The additives also reduce the flash point of the system, thus providing an additional safety factor in treatment and disposal. This property of the additive is expected to be of increasing importance when new air pollution control requirements for treatment units come into effect.

Figure 3.2

Comparison of TCA% Reduction Results for Carbon vs. Rubber Particulate

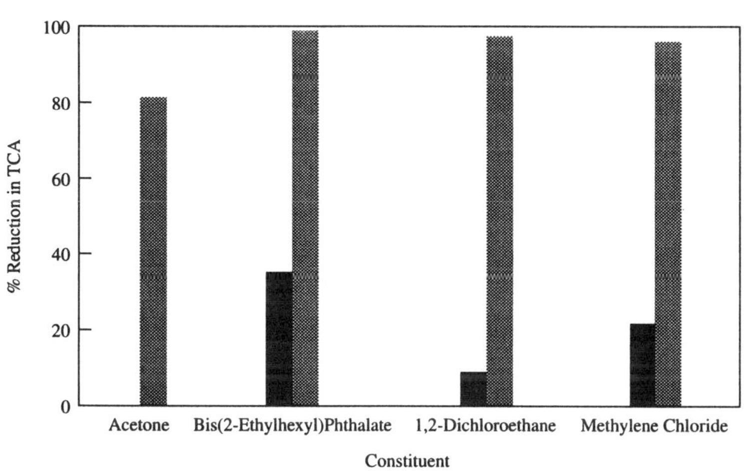

Source: Environmental Technologies Alternatives 1994a

3.1.1.2 Cementitious Solidification/Stabilization

***ProFix*™.** The ProFix process has been commercially applied primarily as a combined fixation agent and filter aid. It is a process employing rice hull ash as the primary agent. However, the vendor has described the results of

Table 3.4
Organic Leaching from ProFix-Treated Waste

	TCLP Leachate (mg/L)	
Constituent	Raw Sludge	Treated Sludge
Methylene Chloride	20.0	<0.25
Chloroform	20.0	2.0
Trichloroethane	2.4	0.56
Toluene	2.1	0.80
Methanol	22.0	5.0
Benzene	30.0	0.76

Table 3.5
Metal Leaching from ProFix-Treated Waste

	TCLP Leachate (mg/L)	
Constituent	Raw Sludge	Treated Sludge
Lead	3.97	0.24
Chromium	7.1	0.05
Cadmium	34.8	0.07
Copper	23.5	0.43
Zinc	158.0	0.27
Nickel	32.1	0.18

several uses as a stabilization system (Tables 3.4 and 3.5). Data cited in the patent (Conner 1992) illustrate the hardening reactions that occur. When a mixture of water and rice hull ash was mixed with sodium hydroxide, no hardening of the paste was observed after 7 days. When calcium chloride was added to the mixture, it hardened in 7 days to >50 kPa (7 psi) bearing strength, and after five months was hard while the sample without the calcium chloride remained a paste. The addition of rice hull ash to a high pH calcium sludge resulted in a very hard product (>430 kPa [62 psi] bearing strength) after 12 days, but no measurable strength in one day. This is compared to a sample treated with sodium silicate solution where the strength of 160 kPa (24 psi) was reached after one day, but only 170 kPa (25 psi) after 12 days, with no additional hardening thereafter. The continued hardening demonstrated by the use of rice hull ash appears to support the chemical theory of this process. The possible advantages of in situ generation of soluble silicate have not been fully explored. Such development should concentrate on the development of long-term properties, i.e., after long curing periods or in long-term leaching procedures, since that is where the advantages will likely lie.

Cement-Slag. The combination of Portland cement and slag for ordinary S/S is not necessarily innovative, but its use for reduction and subsequent stabilization of high-valence species is unusual and deserves inclusion here.

Oak Ridge National Laboratory (ORNL) tested various mixtures of slags in combination with Portland cement and fly ash for the S/S of radioactive wastes containing technetium and nitrates (Gilliam, Dole, and McDaniel 1986). Technetium (^{99}Tc) is more mobile in the higher valence state, Tc^{+7}, than in the +4 oxidation state. Therefore, reduction of Tc^{+7} to Tc^{+4} is desirable in a stabilization process. This can be accomplished by a variety of common reducing agents, such as $FeSO_4$ or Na_2S, but it was believed that blast furnace slag, because of the presence of ferrous iron and sulfur, might accomplish the same purpose at lower cost. The purpose of this project was to test that hypothesis. The waste used in the ORNL test project originated from the treatment of an aqueous effluent, or "raffinate," from uranium recovery at the Portsmouth Gaseous Diffusion Plant in Portsmouth, Ohio. The baseline stabilization composition used in the test program was:

- 38.3% waste sludge,
- 11.7% water,
- 25.0% Type I-II-LA Portland cement, and
- 25.0% fly ash, ASTM Class F.

This mixture yielded treated waste that leached below US EPA primary drinking water standards in the Extraction Procedure Toxicity (EPT) test, and resulted in Leachability Index (LI) values of about 6 when subjected to the ANSI 16.1 Leach Test (ANSI 1986). The filtrate was used for testing the various "fixatives" — blast furnace slag, iron filings, $FeSO_4$, and Na_2S, since it had similar concentrations of Tc and nitrate. The fixative results compared to the treated waste without additives are shown in Table 3.6.

Table 3.6
Effect of Various Additives on Technetium

Constituent Added to Grout	Grout Composition (% by weight)					
	1	2	3	4	5	6
Raw Waste	13.9	13.9	13.9	13.9	30.0	40.0[*]
Water	36.1	36.1	36.1	36.1	20.0	0.0
Cement	25.0	23.2	24.0	24.6	24.6	20.0
Flyash	25.0	23.2	24.0	24.6	24.6	20.0
Iron Filings		3.7				
$FeSO_4$			2.0			
Na_2S				0.9	0.9	
Slag						20.0
ANSI/ANS 16.1 Leachability Index (30-day cure)						
^{99}Tc	7.7	8.1	9.3	10.0	9.4	10.5

[*]Filtrate

The results clearly demonstrate that adding blast furnace slag improves retention of Tc. Six different slag sources were tested, and all gave similar results. The $FeSO_4$ and Na_2S additives gave results similar to those of the slag. Since the LI values are the negative logarithm of the effective diffusion coefficient, the slag additive improved Tc retention by almost three orders of magnitude over the standard cement/fly ash formulation, and by 0.5 to 2.4 orders over the values obtained with other additives. Other data, not shown here, indicated improved nitrate retention also, although only by 1.4 orders of magnitude. The Tc results are largely attributed to reduction of Tc from the +7 to the +4 valence state, reduced porosity and increased tortuosity. The latter effect improved nitrate retention.

The treatment of Cr^{+6} with slag and cement was reported by Rysman de Lockerente (1979), but no leachability data were given. In 1991, Barth et al. (1995) conducted a laboratory treatability study on a sodium dichromate-contaminated soil using three reducing agents to reduce the Cr^{+6} to levels that could be stabilized to within US EPA TCLP limits. Results of the treatability study are summarized in Table 3.7. The agents were sodium metabisulfite and ferrous sulfate, both standard chromium reducing agents, and blast furnace slag. When the reducing agents were used alone, neither sodium metabisulfite nor slag met the requirement, but ferrous sulfate did. However, when the agents were combined with cement in a complete S/S process, both slag and ferrous sulfate gave nearly the same acceptable results. These results clearly show that slag is most effective for chromium reduction in soils when used in combination with Portland cement.

The slag process has been applied only to metal-bearing wastes; however, the processes should have general applicability except where reductive properties would be undesirable, such as in the treatment of certain arsenic-contaminated wastes. Broader testing of blast furnace slag, both as a primary stabilizater and as an additive, would be beneficial. Slag is inexpensive and available in most industrial locations. The cost-effectiveness of slag processes decrease as the distance from the source to the site increases, due to economic competition from locally-obtained waste materials, such as fly ash, kiln dust, etc.

3.1.1.3 In Situ Stabilization and Stabilization/Solidification

Auger-type in situ systems for full-scale projects are new, and process and mechanical improvements are still being developed (Millgard and Kappler 1992). One of the first demonstrations of this technology was in 1988 using the Geo-Con system at a US EPA Superfund Innovative Technology Evaluation (SITE) project in Florida (US EPA 1991c). A small number of full-scale projects have taken place since 1988. Morse and Dennis (1994) described one project in which 1,800 overlapping 2.4 m (8 ft) diameter holes were drilled to stabilize organic contaminated soils with a 10% addition of Portland cement. The stabilized soil had an unconfined compressive strength of 414 kPa (60 psi), permeability of $1 \cdot 10^{-3}$ cm/sec, and leached less than 10 mg/L of polyaromatic hydrocarbons (PAHs).

3.1.2 Design Basis

The information necessary to design an aqueous S/S remediation process or project is basically the same for all aqueous S/S processes. A summary of the important design considerations and necessary data is given in Table 3.8 (Conner 1990; US EPA 1989a).

Table 3.7
Leaching of Treated, Dichromate-Contaminated Soil

| Type of Sample | Concentration of Cr^{+6} and Total Chromium in TCLP Leachates From Treated Soil (mg/L) | | | | | | | |
| | Reducing Agent Alone | | | | Reducing Agent + Cement (Binder/Soil Ratio = 0.2) | | | |
	Cr^{+6}	% Reduction	Total Cr	% Reduction	Cr^{+6}	% Reduction[*]	Total Cr	% Reduction
Untreated Soil	38.0	—	38.5	—	9.65	74	9.3	76
$Na_2S_2O_5$ Treated Soil	20.5	46	20.0	48	11.0	70	9.5	75
$FeSO_4$ Treated Soil	3.65	90	3.3		1.15	97	1.10	97
Slag Treated Soil	30.0	21	28.5		2.05	94	1.8	95

[*]Percent leaching reduction is calculated on the basis of chromium leaching from the untreated raw soil samples.

Table 3.8
Data Input and Considerations for S/S Process Design

Type of Information	Specific Information	Source
Regulating Body	US EPA Region	CD
	State	Permit or CD
Regulatory Framework	CERCLA	FR
	RCRA	CFR
	State Laws	
	Private	RI
	Landbans	FS
	Universal Treatment Standards	ROD
	Site-Specific	
Waste Characteristics	Metal Content	RI
	Metal Speciation	FS
	Organic Content	TS
	Ignitability	RWP
	Corrosivity	Pilot Study
	Reactivity	
	Radioactivity	
	Physical Characteristics	
	General Chemical Characteristics	
Required Stabilized Waste Properties	Strength	FS
	Permeability	TS
	Leachability	RWP
	Durability	
	Transportability	
Site Characteristics	Water Table	RI
	Climate	FS
	Location, Depth and Extent of Waste	ROD
	Soil/Geologic Characteristics	RWP
	Presence of Debris	Site Walk
	Site layout	Pilot Study
	Logistics	
	Utilities	
Operational and Economic Factors	Availability and Cost of Reagents	RI
	Quality and Consistence of Reagents	FS
	Pretreatment Requirements	ROD
	Materials Handling	RWP
	Volume and Weight Increase	Pilot Study
	Future Land Use	
Test Methods	Physical	FS
	Strength	ROD
	Permeability	CD
	Durability	TS
		RWP
	Chemical	
	Leachability	

CD	Consent Decree
CFR	Code of Federal Regulations
FR	Federal Register
FS	Feasibility Study
RI	Remedial Investigation
ROD	Record of Decision
RWP	Remediation Work Plan
TS	Treatability Study

In the case of auger-type, in situ stabilization, some additional information is required since the reagent is normally injected in the waste as a slurry, not in dry form. In this application, it is necessary to test the formulation to determine if it can be mixed and delivered hydraulically within the timeframe and with the type of pumping and injection equipment to be used. Some commonly used S/S reagents, e.g., kiln dusts and quick lime, usually react with water too rapidly to be used in the auger-type system. This limitation should be addressed in the laboratory treatability study so that it does not affect the engineering design considerations.

3.1.3 Design and Equipment Selection

All aqueous S/S processes have the same basic elements (Conner 1990) when used in remediation projects, with certain variations for in situ designs. The scale-up of aqueous S/S systems from bench-scale to field-scale implementation is usually quite straightforward and is well defined. Pilot studies are usually not required in ex-situ treatment except to assess operational problems. A pilot test should always be conducted for in situ S/S applications.

3.1.4 Process Modifications

Aqueous S/S processes and systems are generally quite flexible and can be easily modified to accommodate atypical site conditions, waste composition, reagent variations, and other variables. They can also be adjusted for reasonable changes of scale, either larger or smaller. However, large changes in scale may necessitate a review of the best reagent delivery system. For example, going from a 10,000 m^3 (13,080 yd^3) project to one of 100,000 m^3 (130,800 yd^3) may justify switching from an ex-situ to an in situ method because at large-scale, in situ may be less expensive.

The laboratory treatability study and pilot phases, if conducted, of a remediation project should always include a sensitivity analysis of the effects of changes in waste composition. Wide variations in water content will usually necessitate a change in reagent/waste ratio to maintain consistent final product quality. Variations in metals concentration, particularly in metals speciation, will often require a change in formulation to meet leachability standards. In addition, various compounds including chlorides, fluorides, sulfates, phosphates, zinc, lead phenols, and many organic compounds can interfere with cement hydration reactions and impede curing of the solidified product. Variations in organic content can affect both leachability, and the immediate and final physical properties of the treated waste. Careful review of the Remedial Investigation and Feasibility Study (RI/FS) and various treatability studies can point out potential problems from waste composition variations, and preclude surprises later on.

3.1.5 Pretreatment Processes

Pretreatment processes that have sometimes been necessary prior to aqueous stabilization or S/S are listed below.

- Cyanide destruction

- Metal chelate destruction

- Cr^{+6} and Tc^{+7} reduction

- As^{+3} oxidation

- Removal of volatile organics by steam or hot air stripping, or thermal desorption

- High-level organic removal by soil washing, solvent extraction

- Destruction of organics by high temperature incineration

- Biological destruction of organics

- Debris size reduction or removal

- Size reduction of certain hard, porous materials

Treatment of inorganics by stabilization or S/S can require additional processing steps. These operations, whether classified as pretreatment, step-wise stabilization or S/S, or cotreatment, can be done with conventional technology, although improved innovative technology might also be available. High levels of organic contaminants of concern, are usually removed or destroyed in a separate pretreatment operation.

Cyanides, chelates, and metals pre- or cotreatment with S/S have been used conventionally at RCRA Treatment, Storage and Disposal Facilities (TSDFs) since the LDRs went into effect in 1988 and earlier at some other central treatment operations (Conner 1990). Cyanide destruction is typically done with alkaline chlorination for amenable cyanides (Conner 1990) or with more powerful oxidants for the more refractory cyanides and for metal chelates (Diel, Kuchynka, and Borchert 1995). Chromium reduction along with S/S was described by Conner (1990; 1991), arsenic oxidation by Lear and Conner (1992), and technetium reduction by Gilliam, Dole, and McDaniel (1986).

Organics can be removed or destroyed in situ by steam or hot-air sparging, chemical oxidation (Siegrist et al. 1992) or soil flushing (US EPA 1988; US EPA 1991b). Likewise organics can be removed or destroyed ex-situ with thermal desorption (Lighty et al. 1993) or incineration (Magee et al. 1994). Cyanide destruction and metal reduction/oxidation should be feasible in situ and ex-situ, but, no full-scale treatment results have been reported.

Debris, a frequent problem in remedial projects, can be handled in two ways: (1) removal by screening or other physical methods, or (2) by grinding, shredding, or similar size reduction techniques. The primary requirement for debris pretreatment stems from limitations on the size or type of debris that the S/S mixing method can accommodate. If the material is porous and contains hazardous constituents that slowly diffuse out to the particle surfaces after S/S has been completed, size reduction might be necessary (Wilson and Clarke 1994). An example is hexavalent chromium in certain waste refractories. By decreasing the diffusion path length and increasing the surface area, particle size reduction allows the stabilization reactions to go to completion during the mixing stage.

3.1.6 Posttreatment Processes

Wastes successfully treated by aqueous S/S processes do not normally require any posttreatment. These wastes are usually disposed by landfilling, or in the case of in situ treatment they are left in place. The landfill design and/or closure plan and post-closure monitoring address possible releases to the environment or dangers to human health. The only posttreatment control required in some instances is to limit particulate and VOC air emissions and odor until the waste is landfilled and/or covered.

3.1.7 Process Instrumentation and Controls

Since the final properties of aqueous S/S products are usually not developed in the process system itself, monitoring these properties in real time is not possible. Quality control is exercised by relating the S/S formulation to the waste properties in the laboratory treatability and pilot studies, and subsequently controlling the reagent and water addition during full-scale processing. Instrumentation is primarily for the purpose of controlling waste, reagent, and water feed rates. The instrumentation and controls used during ex-situ processing are standard in chemical processing and concrete formulation. Most waste properties cannot be measured in real time, but some can be and are used to "fingerprint" the incoming waste to detect gross variations. These include continuous pH, water content, and density measurements. If a proper sensitivity analysis was performed in the treatability and pilot studies, variations in these properties can often be used to change reagent and water feed rates in real time, even in continuous systems. At least one such system successfully used a computerized control system for a continuous S/S plant (Bounini 1990).

For in situ, auger-type systems, such fingerprinting is not feasible. However, reagent and water feeds can be controlled accurately.

3.1.8 Safety Issues

Three kinds of safety concerns exist when using aqueous S/S — those associated with the waste, the reagents, and the equipment. All three should be covered in the project health and safety plan. Concerns with the waste have usually been thoroughly determined and accounted for during the RI/FS, project design, and implementation planning. Most of the reagents used are not toxic or hazardous in and of themselves, but can pose hazards when reacting with water and/or the waste; some regents do require special handling for safety reasons. These hazards fall mostly into three categories: inhalation of fine particulates, heat generated by exothermic reactions, and evolution of gases. In some situations, sufficient heat may be generated to cause burns and even pose fire and explosion hazards. Ammonia is the gas most likely to be generated by reaction with the alkaline reagents commonly used in aqueous S/S. These potential hazards must be addressed in the health and safety plan by providing proper ventilation, personal protective equipment, and personnel training.

3.1.9 Specification Development

Most specifications used to procure vendors, equipment, and facilities in aqueous S/S projects are similar to those used in chemical processing and construction, modified to meet the needs of the project. The only unusual requirements pertain to reagent quality, since some of the reagents used are themselves waste products, e.g., kiln dust and fly ash. The quality of Portland cement and lime is well-defined by the ASTM and government specifications. Kiln dust and fly ash are best defined by the content of free CaO, since that is the principal reactive ingredient; wide variations in quality occur with these reagents. Other additives are usually commercially-available chemicals and other products, the quality of which is controlled by the supplier.

3.1.10 Cost Data

All of the aqueous processes described, with the exception of phosphate and organic stabilization, use reagents that are conventional and commercially available in North America in most locations at competitive prices. Phosphate processes most frequently use phosphoric acid, and can often use the cheaper agricultural grade. Organic stabilization processes may use virgin or recycled carbon, rubber particulate, organo-clays, or more exotic additives. The latter additives are nearly an order of magnitude more costly than the basic binders, but are normally used at much smaller concentrations. Typical delivered prices for reagents are shown in Table 3.9.

Table 3.9
Reagent Costs

Reagent	1996 Price ($/ton)
Portland Cement	60-80
Blast Furnace Slag	10-20
Quicklime (untreated)	50-90
Rice Hull Ash	about 400
Phosphoric Acid	300-600
Activated Carbon (virgin)	1000
Activated Carbon (recycled)	500
Rubber Particulate	300-400
Organo-Clays	500-1,000
Fly Ash	15-40
Cement Kiln Dust	20-40

Except for a few phosphate process cases, all of the wet processes use much the same equipment and processing techniques; the primary difference is between in situ and ex-situ delivery systems. Therefore, variations in the commercial cost of aqueous S/S for different processes is determined primarily by reagent costs for each delivery method and can be easily assessed and compared on this basis. In the case of the phosphate process, the application technique can sometimes be simple spray application of the phosphoric acid or phosphate solution while the waste is moving through a conveying system. In this instance, the processing cost is substantially less.

Since aqueous S/S is, in general, a proven commercial technology (or group of technologies), the overall costs are well-established. For example, costs are shown in Table 3.10 for one S/S scenario (Conner 1992) involving a large-scale remediation of a low-hazard, lead-contaminated soil at a "dry" site in an industrial plant property. Of the total price for the remediation cost of about $83/tonne ($75/ton), $55/tonne ($50/ton) is for the stabilization

Table 3.10
A Typical Example of Aqueous, Ex-Situ S/S
Remediation Costs in 1990 with On-Site Landfill*

Cost Item	Case "A"
Waste Type	Soil
Contaminant/Level	Lead/400 mg/kg
Waste Quantity	50,000 yd^3
Transportation	—
Mobilization/Demobilization	$25,000
Treatability Study and QA/QC	$30,000
Excavation	$250,000
Stabilization	$2,500,000
Landfill	$1,000,000
Total Cost	**$3,805,000**
Approximate Total Cost per yd^3	**$76**

*These costs do not include typical oversight and monitoring.

operation. Of this \$55/tonne (\$50/ton), the typical reagent cost for a cement-based process would be about \$22/tonne (\$20/ton), or about 40% of the overall stabilization cost. This ratio is common for conventional processes on large-scale projects.

The US EPA has published a guide for documenting remediation costs for various technologies (US EPA 1995a) which provides different levels of cost categorization. For portable treatment units typically used in remediation, the fifth level standardized work breakdown structure (WBS) applicable to aqueous S/S is given in Table 3.11. Comparison of Tables 3.10 and 3.11 indicates that some items in the US EPA WBS are usually combined in published S/S cost breakdowns. Also, the US EPA format does not specifically include QA/QC costs, which are included in Table 3.10.

For in situ S/S (and stabilization alone), solids preparation and handling (WBS activity -01) is not required. The other activities are similar, but the costs may be different, especially in the operation phase.

Table 3.11

Fifth Level Work Breakdown Structure Cost Elements for Aqueous S/S

Interagency WBS #33 XX XX 01–	Case "A"
01	Solids Preparation and Handling — Includes loading/unloading, screening, grinding, pulverizing, mixing, moisture control, and placement/disposal.
04	Pads/Foundations/Spill Control — Includes materials and construction of facilities.
05	Mobilization/Setup — Includes activities needed to prepare for startup.
06	Startup/Testing/Permits — Includes activities needed to begin operation.
07	Training — Includes training needed to operate equipment.
08	Operation (Short-term — Up to 3 years) — Includes bulk chemicals/raw materials, fuel and utility usage, and maintenance and repair.
09	Operation (Long-term — Over 3 years) — Includes bulk chemicals/raw materials, fuel and utility usage, and maintenance and repair.
10	Cost of Ownership — Includes amortization, leasing, profit, and other fees not addressed elsewhere.
11	Dismantling — Includes activities needed prior to demobilization.
12	Demobilization — Includes removal of unit.

3.1.11 Design Validation

The physical/mechanical design of the aqueous S/S system is essentially the same for all of the individual technologies, varying primarily in whether the process is operated in situ or ex-situ. Because this design is well proven in many commercial applications, and scale-up and other variables are well established, validation is normally not required. The only mechanical design considerations for individual projects are the choice of type, size, and robustness of material handling equipment, particularly the mixer. These choices are almost always among standard, off-the shelf equipment. The proper application of the particular chemistry involved, however, must be validated for the waste type and operating scenario under consideration. Chemical design validation is done through bench- and pilot-scale testing on actual waste materials In the pilot phase, either scaled-down or full-scale commercial equipment may be used. In the case of ex-situ S/S, the pilot phase is often omitted, but pilot testing should never be neglected when in situ treatment is to be done because of the greater uncertainties encountered in the latter type of operation. A number of laboratories, consulting and engineering firms, and S/S vendors are available for evaluating various aqueous S/S technologies and for choosing the best physical/mechanical design for the operating system.

3.1.12 Permitting Requirements

The types of information required by regulatory agencies to gain approval for aqueous S/S remediation projects were discussed in Section 3.1.2 and listed in Table 3.8. Generally, this information is available from the combination of RI/FS, Record of Decision (ROD), Consent Decree, treatability study, and pilot study. In addition, the regulatory agencies will require a specific project work plan, health and safety plan, quality assurance project plan (QAPP), and financial and liability information. Depending on the project, the agency may also require closure and post-closure maintenance and monitoring plans.

The primary purpose of these requirements is to ensure that the project will be carried out as planned, that the treated waste will meet the performance specifications, and that human health and the environment will not be unacceptably affected. Once the plan is approved, specific environmental permits may be required, depending on the state in which the project is located. These permits are typically issued by the state and/or local environmental regulatory agencies. They can include air permits, groundwater treatment permits, hazardous waste treatment and disposal permits, and/or building/zoning permits. Each project is site-specific.

3.1.13 Performance Measures

There are two aspects of performance: process performance and product performance. The latter subject is covered in Section 4.1.5, Quality Assurance/Quality Control (QA/QC). Process controls, of course, ultimately affect product QA/QC, but the methods and parameters are different. The controls used in aqueous S/S processes were discussed in Section 3.1.7.

3.1.14 Design Checklist

3.1.14.1 Ex-Situ Aqueous Stabilization/Solidification

1. Characterize the waste stream to determine the following parameters: homogeneity; presence of debris; moisture content; particle size; pH; total alkalinity/acidity; reactivity, especially at high pH; and major constituent composition (metals and organics, including nonhazardous organics, such as oil and grease).

2. Determine if pretreatment of waste is required. Examples are chromium and technicium reduction, particle size reduction, screening, pH neutralization, and dewatering.

3. Determine the product property requirements.

4. Complete a proper treatability study.

5. Determine if there is any anticipated end use of the product or the site. Determine final disposition.

6. Ascertain if there are regulatory requirements, such as permits, etc. and initiate the permitting process.

7. Evaluate site conditions, such as working space, access for equipment and reagent delivery, weather, groundwater infiltration, soil type and conditions, and underground utilities.

8. Determine if emission controls are required for volatile organics, NH_3, sulfur oxides, etc.

9. Determine the throughput requirements.

10. Complete pilot study, if necessary.

11. Prepare a QA/QC (QAPP) for acceptance by the regulatory agencies.

3.1.14.2 In Situ Aqueous Stabilization/Solidification

1. Characterize the waste site/stream to determine the following parameters: homogeneity; presence of debris; moisture content; particle size; pH; total alkalinity/acidity; reactivity, especially at high pH; and major constituent composition (metals and organics, including nonhazardous organics, such as oil and grease).

2. Determine if pretreatment of the waste is required. Examples are chromium and technicium reduction and VOC stripping.

3. Determine the product property requirements, such as strength, permeability, leachability, durability, and total constituent analysis (organics).

4. Complete in situ treatability study.

5. Determine if there is any anticipated end use of the product or the site.

6. Ascertain if there are regulatory requirements, such as permits, etc. and initiate the permitting process.

7. Evaluate site conditions for working space; weather; access for equipment and reagent delivery; groundwater infiltration; soil type and conditions; soil bearing strength (for support of heavy equipment); presence of debris; presence of "hot spots"; site geology; and underground utilities.

8. Determine if emission controls are required for volatile organics, NH_3, sulfur oxides, etc. If so, select a proper size Air Pollution Control system.

9. Determine treatment rate requirements.

10. Complete pilot study.

11. Prepare a QA/QC (QAAP) for acceptance by the regulatory agencies.

3.2 Polymer Stabilization/Solidification

3.2.1 Remediation Goals

When applied in ex-situ S/S, polymer technologies can provide improved mechanical integrity, durability, and leachability in the treated waste compared with conventional S/S technologies (Neilson and Colombo 1982). Likewise, in situ applications of polymer S/S for treatment of buried waste and contaminated soils improve performance of the stabilized product (Heiser 1995). Difficulties in processing "problem" wastes with conventional hydraulic cement grout and widespread failures of hydraulic cement waste form products have been well documented (Neilson et al. 1983; Place 1990; NRC 1991a; Lomenick 1992). In addition, conventional, hydraulic-cement grout waste forms are relatively porous and therefore have relatively high leach rates, especially for some contaminants, such as cesium, that are not well-bound chemically (Colombo and Neilson 1979).

In contrast, compatibility in processing a wide range of wastes and durability under anticipated disposal conditions has been established for several polymer binders. The ability to withstand degradation from saturated soil conditions, freeze-thaw cycling, microbial degradation, and high radiation environments, has been confirmed for polyethylene (Kalb, Heiser, and Colombo 1991a), vinyl ester-styrene (Dow Chemical Co. 1978), other thermosetting polymers (Heiser and Milian 1994), and sulfur polymer (Kalb et al. 1991). Leaching rates for waste that is microencapsulated in polymer waste forms vary according to the type of polymer, the characteristics of the waste (e.g., solubility, particle size), and the quantity of waste encapsulated (waste loading). Typically, polymer leaching rates are about two orders of magnitude lower than those of conventional cement grout waste forms with equivalent or lower waste loadings. For example, polyethylene encapsulation of nitrate salt wastes, which are highly soluble and thus, susceptible to leaching, reduced long-term leachability by a factor of 75 compared with conventional cement grout (Fuhrmann and Kalb 1993).

Since no chemical reactions are required to solidify the final waste form and solidification is assured on cooling, thermoplastic polymer processes are inherently more reliable than conventional hydraulic cement grout systems. Regulatory acceptance of ex-situ polymer S/S technologies is being established. For example, the Nuclear Regulatory

Commission (NRC) has approved a topical report on the use of vinyl ester-styrene for solidifying commercial low-level radioactive waste streams (Dow Chemical Co. 1978). Other polymers, including polyethylene and sulfur polymer, have passed required test criteria for NRC approval, but have yet to be submitted by a commercial vendor for formal licensing. The US EPA has identified polymer macroencapsulation of radioactive lead solids as the recommended BDAT (US EPA 1989c). Considering that polymer S/S is a simple technology that is easily understood, widespread support from the public is anticipated.

Following successful technology development at Brookhaven National Laboratory and Rocky Flats Environmental Technology Site, polymer macroencapsulation of radioactive lead has been commercialized by Envirocare of Utah, Inc. Under a contract with the DOE, Envirocare, the only commercial, licensed mixed waste disposal facility in the U.S., is completing the treatment and disposal of 227,000 kg (500,000 lb) of mixed waste lead shipped from various sites throughout the DOE complex. In addition to radioactive lead, Envirocare is planning to treat mixed waste debris by polymer macroencapsulation and a range of mixed waste solids (e.g., salt and ash) by polymer microcapsulation. The State of Utah Department of Environmental Quality has issued an operating permit for the macroencapsulation process and is currently reviewing Envirocare's permit modification request to commercialize microencapsulation.

Application of thermosetting polymers for in situ treatment of buried waste (e.g., drums and debris) is currently under development in a collaboration between Idaho National Engineering Laboratory, Brookhaven National Laboratory, and Sandia National Laboratory (Heiser 1995). The objective of this effort is to stabilize buried waste and reduce dispersion during planned removal actions. Other potential applications of in situ polymer S/S are treatment of contaminated soils and allowing the stabilized product to be left in place. In a related effort that investigated the use of polymers for in situ barrier walls around contaminated sites and leaking tanks, extensive data on performance of polymer grout/soil combinations indicate that this technique can result in a stable, durable product. Testing of polymer grout/soil combinations has included hydraulic conductivity; compressive strength; flexural strength; splitting tensile strength; water immersion; acid, alkaline, and solvent resistance; wet-dry cycling; chloride diffusivity; thermal cycling; and irradiation stability (Heiser and Milian 1994).

Regulatory issues associated with in situ injection of polymers in relation to in situ polymer barriers was also investigated (Siskind and Heiser 1993). The US EPA indicated that injection of polymers was essentially a construction operation and did not envision any regulatory problems as long as the final product is inert. For in situ S/S applications, further performance testing, such as leaching, of the final products would be required to ensure compliance with federal, state, and local environmental regulations.

3.2.2 Design Basis

As with other potential treatment technologies, polymer processes require thorough characterization of the waste to enable technically-sound and cost-effective selection of a polymer S/S system. Waste properties, including types and concentrations of contaminants, chemical constituents, moisture content, and particle size, are important. Location and condition of the waste materials (e.g., contaminated soil vs. storage of sludge in buried drums), as well as the total volume requiring treatment, play a significant role in technology selection. Treatability studies which determine compatibility of the waste and treatment systems, examine integrity and durability of the final waste forms, and demonstrate compliance with applicable regulatory performance criteria are also key elements in the selection and design of polymer systems. Specific design issues for each polymer technology are discussed below.

3.2.2.1 Polyethylene Encapsulation

Since polyethylene can be used for either microencapsulation or macroencapsulation, waste characterization data, such as particle size and contaminants, are needed to determine which technology approach will be used. For mixed waste debris or contaminated solids with particles >60 mm (2.4 in.), macroencapsulation can be used for cost-effective treatment, according to US EPA regulations (US EPA 1992b). For aqueous solutions/dissolved solids, sludges, resins, ash, and other wastes with relatively small particle size, polyethylene microencapsulation provides improved performance in the areas of mechanical integrity, durability, and leachability (Kalb and Colombo 1997). Since polyethylene melts at 120°C (248°F), residual moisture driven off during processing can cause the formation of voids due to gaseous entrapment within the melt. Therefore, moisture content can impact selection of optimal processing technology and/or the need for pretreatment to remove residual moisture.

Likewise, if significant concentrations (e.g., >1% by weight) of volatile organics are present in the waste, they can be liberated during processing. In such cases, the waste will require thermal treatment to destroy the organic constituents (thermal destruction) or capture of the organics in the offgas (thermal desorption) for further treatment.

3.2.2.2 Sulfur Polymer Cement Encapsulation

Because sulfur polymer cement (SPC) is a thermoplastic process and does not require a chemical reaction for solidification, it is also applicable for the treatment of a wide variety of wastes. It has been shown to be effective for incinerator ash, aqueous concentrates (sulfate, borate, and chloride salts), sludge, and debris (Kalb and Colombo 1985a; Kalb, Heiser, and Colombo 1991b; Van Dalen and Rijpkema 1989; Darnell 1991). It is not recommended for treatment of ion exchange resin wastes, because it is unable to withstand swelling stresses that occur when resins are rehydrated. It is also unsuitable for nitrate wastes, because of potential dangerous reactions between sulfur, nitrate, and trace organics (Kalb and Colombo 1985a). Strong alkali (over 10%), strong oxidizing agents, aromatic or chlorinated hydrocarbons, oxygenated solvents, carbon disulfide, bromoform, and other sulfur-dissolving solvents might also have a deleterious effect on SPC waste forms. Additional durability testing is recommended if these conditions are expected.

There are few data available on the long-term durability of SPC, hence short-term tests have been used to project durability. The current test methods for biodegradation, ASTM G-21 and G-22, are not sufficient for sulfur final waste forms, and the main concerns about SPC durability are biodegradation and the ability of bacteria to oxidize elemental sulfur to sulfate over a broad pH range (Mattus and Mattus 1994).

Sulfur polymer cement can be used in microencapsulation or macroencapsulation applications, hence the need to determine the waste particle size. Its low melt viscosity makes it potentially well-suited for macroencapsulation of debris waste, but this application has not been demonstrated. As with other thermoplastic technologies, SPC is processed at temperatures above the vaporization temperature of water, so that pretreatment to drive off excess residual moisture is recommended. However, since it is a batch process, it is more tolerant of moisture contained in the waste than is the polyethylene process. Small quantities of residual moisture are easily driven off during batch mixing. Re-melt capability allows the final waste form to be easily remediated if unforeseen degradation occurs or if more stringent future disposal requirements dictate.

3.2.2.3 In Situ Polymer Stabilization/Solidification

Concentration and density of contaminants in soils are significant issues when considering this remediation alternative. For soils in which contaminants are highly concentrated and/or localized to a small discrete area, in situ injection of polymers for stabilization might be feasible. However, high concentrations of contaminants in soils could exceed the capabilities of in situ polymer S/S techniques to reduce contaminant migration to acceptable levels.

Performance assessment (PA) modeling of the site (under baseline conditions prior to treatment and following polymer S/S) is useful in predicting the potential success of in situ processes. In order to conduct credible PA modeling, however, site-specific data on soil characteristics such as hydraulic flow and the sorption coefficient, K_d, are required. These data are generated through field measurements and bench-scale testing using actual soil and groundwater samples, respectively. To accurately predict leachability reduction following polymer S/S, bench-scale and/or field testing of polymerized soils to determine permeabilities and leach rates, are required. In situ polymer S/S might not be effective for soils contaminated with high concentrations of organics, so extensive analyses of constituents, including organics, inorganics, and radionuclides, are needed. Considering the relatively high cost of thermosetting polymer materials, large volumes of contaminated soils might not be cost-effectively treated by in situ injection. Therefore, field monitoring data on the concentration and geographical density of contaminants are required.

3.2.3 Design and Equipment Selection

Design and selection of processing equipment for polymer S/S depend on the type of polymer (e.g., thermoplastic polymer such as polyethylene or thermosetting polymer such as polyester styrene) and the manner in which the final waste form will be produced (ex-situ vs. in situ, microencapsulation vs. macroencapsulation). In most cases, these technologies can be implemented using readily available, "off-the-shelf" components, thereby lowering capital costs and reducing time required for implementation. Engineering requirements are primarily in the areas of sizing, specifying, configuring, and modifying equipment to assemble components into an integrated processing system.

Details on system design and equipment selection are available in the literature for some technologies (Patel, Lageraaen, and Kalb 1995), but as with other emerging technologies, the engineer must often rely on prior knowledge and engineering principles to adapt the necessary equipment required to fully integrate the process. The following sections summarize available design and equipment selection data, to provide a general background for engineers and site administrators. As with other technologies, final equipment selection must also account for waste- and site-specific conditions.

3.2.3.1 Polyethylene Encapsulation

MICROENCAPSULATION. Microencapsulation of waste in polyethylene involves processing the thermoplastic binder and waste materials into a waste form product by heating and mixing both materials into a homogeneous molten mixture. Cooling of the melt creates a solid monolithic final waste form in which contaminants have been completely surrounded by a polymer matrix. Heating and mixing requirements for successful microencapsulation of waste in polyethylene can be met using proven technologies available in various types of commercial equipment. Processing techniques for thermoplastic materials, including polyethylene, are well-established within the plastics industry. Extruders and mixers are available in a broad range of designs for the manufacture of consumer and commercial products as well as for compounding applications. Compounding (mixing additives, such as stabilizers and/or colorants, with polymers) is analogous to microencapsulation of waste in thermoplastic polymers. The majority of commercial polymer processing has historically been accomplished by extrusion, a process that has been used successfully for over 60 years. In the extrusion process, materials are heated to a molten state, mixed, forced through a die under pressure into a mold, and allowed to cool, forming a solid product. Various types of extruders are available (e.g., single-screw, twin-screw) and each can be custom engineered by varying design parameters, such as barrel diameter, length/diameter ratio, screw design, heating and cooling systems, etc. In addition, numerous alternative techniques, such as continuous, batch, and thermokinetic mixers, thin film evaporators, and screwless extruders have been used to process polymers. Each of these devices is reviewed here, with an emphasis on those systems most applicable for microencapsulation of waste.

Extruders — General. Screw extruders operate by converting a flow of thermoplastic material and additives into a well-mixed continuous melt stream. Extruder design, as shown in Figure 3.3, consists of a rotating

Figure 3.3
Schematic of Typical Plastics Extruder

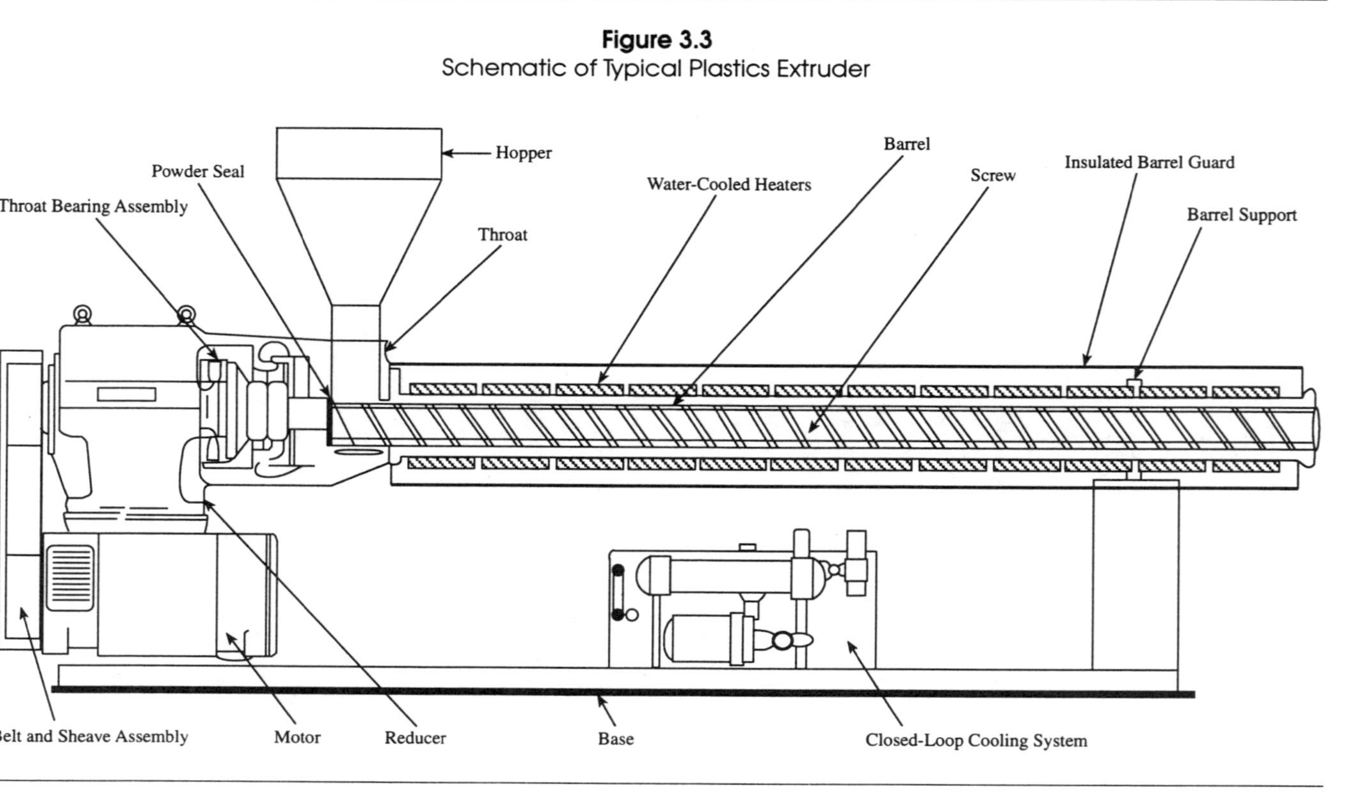

Source: Frados 1985

screw(s) within a barrel, forming a channel between the root of the screw and the interior barrel surface. Material residing within this channel is mixed, melted, and conveyed by helical flights on the rotating screw.

The process can be divided into three zones common to all screw configurations. Beginning at the feed hopper, thermoplastic resins (in the form of beads, pellets, or powder) are fed directly into the feed throat. Additives, including waste to be encapsulated, can be fed with the polymer or introduced upstream by a "crammer" feeder. This feed zone contains deep channels and long screw flights that initiate the process by filling the channel, building a pressure gradient, and conveying the unmelted ingredients through the barrel. Next, the material enters the transition zone where external thermal energy combines with frictional heat to melt the polymer. External heat is usually provided by a series of electric resistance heaters. Cooling, necessitated by the buildup of frictional heat, is provided by cooling fans or by circulating liquid coolant. The temperature is precisely controlled by solid-state proportional-integral-derivative controllers. The reduced channel volume in the transition zone compresses the unmelted material, eliminates air pockets, and further increases the pressure. The ingredients are mixed and plasticized by the intense shear generated by the motion of the screw. Some extruders have a venting zone about two-thirds of the way down the barrel. An increase in channel volume induces a pressure drop, enabling vacuum to be applied at the vent to remove volatiles remaining in the melt. Finally, a sharp decrease in the channel volume of the metering or pumping zone (located at the end of the screw) further compresses the melt and increases the melt pressure in order to pump the material through an output die. Extruder output dies are designed to meet final product configuration requirements, but for the production of final waste forms, complex die configurations are not required.

The motion of the screw advances the material throughout the process and applies necessary shear forces to blend the polymer and additive. Screw extruders have limited ability to remove residual moisture or other volatile gases in excess of 5% (by weight) of the feed material, even when equipped with vacuum venting. Volatile gases that remain trapped within the melt cause foaming, undesirable voids, and reduction in final waste form density. This reduces process efficiency, degrades mechanical integrity, and increases leachability of the final waste form product. Pretreatment processes discussed in Section 3.2.5 (e.g., drying), might be used to make particular waste streams more amenable to extrusion processing.

Screw extruders are primarily classified by the number of screws they contain. The two most common screw extruders, single-screw and twin-screw, have been used for waste encapsulation and are discussed here. Other variations include two-stage compounding extruders, reciprocating-screw kneader, and concentric-screw extruders.

Single-Screw Extruders. Single-screw extruders are well-proven in the plastics industry for processing virgin and recycled thermoplastic polymers. They have also been used for certain compounding applications such as the mixing of fillers, colorants, and additives with a wide variety of thermoplastics. This broad applicability is achieved with various screw designs. The choice of screw type depends largely on the physical properties (liquid/solid compatibility, bulk density, specific gravity, etc.) of the materials being processed, as well as the level of mixing and process rates desired. Basic screw designs include feed, transition, and metering sections with uniform screw flights. Advancements in screw configuration have led to the incorporation of second flights, interrupted flights, high-shear barriers, and various types of mixing sections to improve performance. Every screw design, however, is a compromise between productivity, melt temperature, degree of mixing, and output uniformity. Examples of some generic, single-screw types are shown in Figure 3.4.

The basic metering screw in a standard single-screw design encompasses the three zones described above. It is routinely used for the extrusion of polyvinyl chloride compounds, fiber, and other fillers and has sufficient mixing capabilities for many applications. Mixing, however, can be improved by the *Maddock mixing screw* that uses a close-clearance mixing section to deliver added shear. Higher melt temperatures normally associated with increased shear are controlled by placing the mixing section in an area of the barrel subject to heating and cooling. This is readily accomplished since most extruders are equipped with temperature controllers along the entire length of the barrel. An alternative mixing mechanism is provided by the *pin mixing screw* which contains multiple rows of pins that break channel circulatory flow patterns to enhance blending without significantly increasing shear. The *barrier screw* design improves melting by separating melt and solids channels with offset barrier flights. This results in increased output and/or reduced melt temperatures and improved pressure stability. The Maddock mixing section can also be integrated with the barrier mixing screw. Another design option is the *two-stage venting screw*. In this option,

Figure 3.4
Various Screw Types Available for Single-Screw Extruders

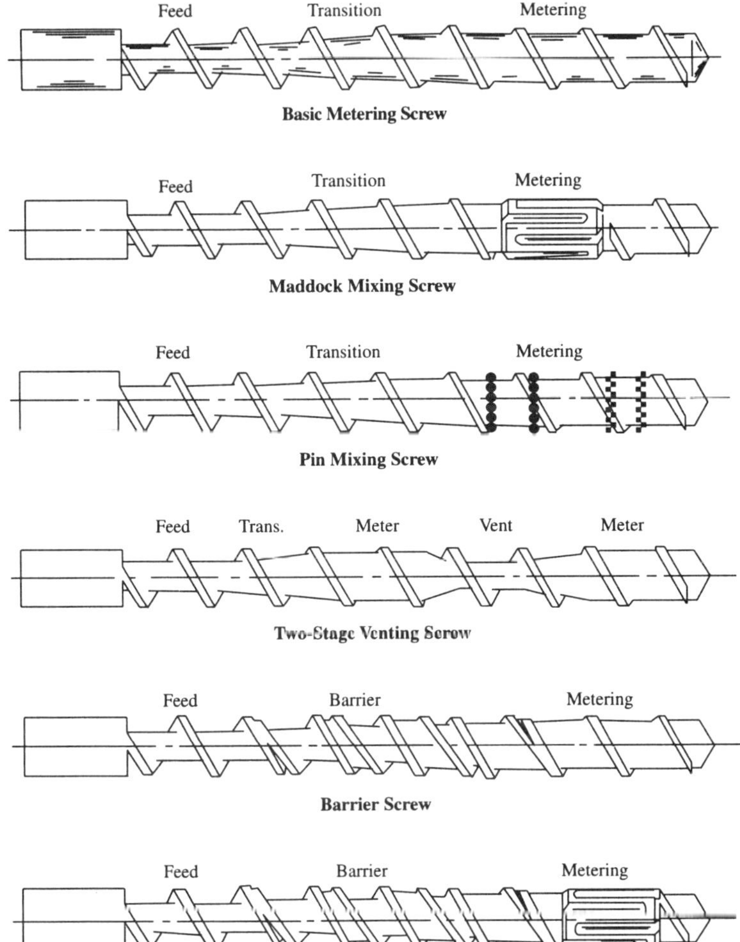

a feed, transition, and metering section in the first stage dumps into a deep-channel area followed by a recompression and pumping zone within the second stage. Volatile gases can be vented through vent ports within the let-down section. Additional variables that control product quality are screw flight pitch angles, length-to-diameter ratio, and operational controls. Furthermore, barrel designs can be modified to include grooved or tapered sections to enhance throughput.

Years of proven application within industry make single-screw extruders a dependable technology that yields advantages in its simplicity, operating ease, low maintenance, and great versatility at lower operating costs when compared to other competing processes. An integrated, full-scale polyethylene encapsulation process based on single-screw extrusion has been successfully demonstrated (see description in Chapter 5).

Waste particle size can limit the use of single-screw extruders. Development work at Brookhaven National Laboratory has indicated that single-screw processing can successfully microencapsulate waste materials with mean particle sizes ranging from 50 to 3,000 μm (2 to 118 mils). Smaller mean particle sizes are difficult to mix with viscous thermoplastics and larger particle sizes require size reduction prior to feeding. Recent advances in screw designs, sophisticated control systems, and modified feed mechanisms have minimized limitations associated with single-screw extrusion. New developments linking single-screw extrusion with pretreatment by thermokinetic mixing promise to reduce or eliminate particle size constraints, while improving overall mixing.

Twin-Screw Extruders. Twin-screw extruders, designed with two screws placed side-by-side, have proven versatile for handling difficult compounding jobs, such as glass-fiber, high-loading fillers and heterogeneous plastics. Screw arrangement can be tailored to meet distinct processing requirements, allowing improved control of critical operating parameters, such as residence time, degree of shearing, and processing temperature. In addition, intermeshing screw flights deliver a unique forced-conveying or pumping property that broadens the conveying capabilities of this type of extruder. As a result, the twin-screw extruder can function like a positive screw-type pump to handle difficult-to-feed materials. Options in screw configuration can further expand versatility through the addition of intense mixing and shearing elements, venting zones, or a variety of process-specific devices. As shown in Figure 3.5, twin-screws are normally categorized as either co-rotating or counter-rotating, with further classification of intermeshing and non-intermeshing designs. A new line of reversible intermeshing twin-screw machines can operate in both co-rotating and counter-rotating modes.

Figure 3.5
Types of Screw Configurations for Twin-Screw Extruders

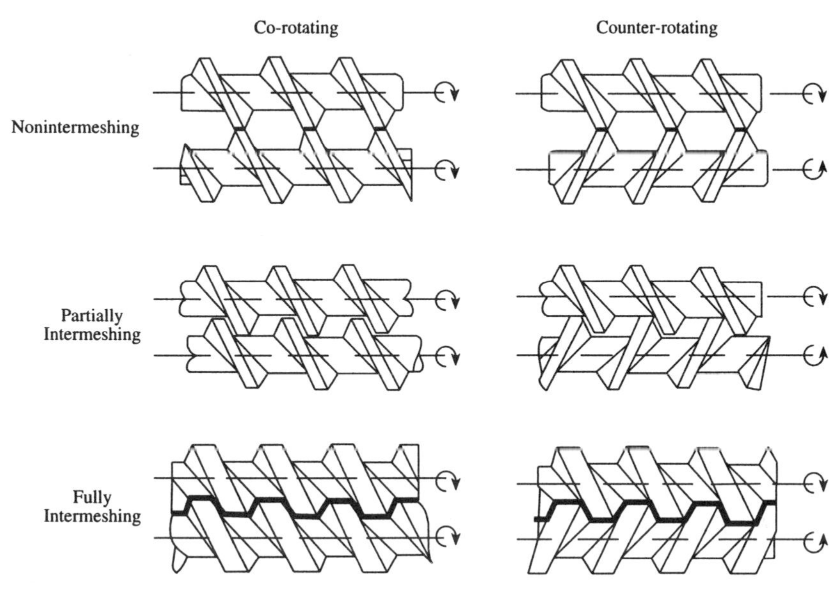

Source: Frados 1985

Co-rotating screws are commonly used in processing nylon, thermoplastic polyester, and polypropylene. In this configuration, two screws rotate in the same direction, deflecting the ingredients on a figure eight pattern around both screws resulting in an excellent exchange of material. At the same time, small uniform clearances between intermeshing screw flights yield a uniform residence time for each melt particle while also helping to eliminate dead spots. Co-rotating screws, however, are limited to conveying material through drag flow mechanics, whereas the counter-rotating intermeshing design operates as a positive-screw type pump, thereby broadening its range of application. A counter-rotating intermeshing arrangement can also provide greater control of mixing, shearing, and conveying properties by regulating the amount of clearance between the screws. For example, narrowing the gap generates higher shear as material is forced through a smaller opening but lowers output with less area for longitudinal transport. Non-intermeshing counter-rotating screws are open both lengthwise and crosswise to promote generous mixing of material between screws, as well as conveying material at higher outputs. However, this "open" design limits the control over and degree of shearing the screw can impart.

Twin-screw extruders were originally adapted for waste encapsulation using bitumen in the mid-1970s (Werner and Pfleiderer Corp. 1976). A production-scale process was installed to process low-level radioactive waste generated at the Pallisades Nuclear Power Plant in Michigan. More recently, development efforts at Rocky Flats Plant have successfully used twin-screw extrusion for microencapsulating wastes in polyethylene. The process has been demonstrated at laboratory-scale for both surrogate and actual mixed wastes (Faucette et al. 1994). Some advantages claimed over single-screw extruders include greater versatility for compounding "difficult-to-mix" materials and better control of shear and temperature parameters. In addition, the gear-pump effect makes it possible to accept difficult-to-feed material as well as allowing for multiple downstream feed zones which can eliminate the need for pre-blending. Material pretreatment requirements are similar to those of single-screw extruders, but due to improved dispersive mixing in twin-screw extruders, a broader particle size range may be tolerated. Disadvantages of twin-screw extruders compared with single-screw are higher capital costs and higher operating costs (due to more frequent and costly maintenance requirements) and more complicated operations.

Thermokinetic Mixers. The major difference between thermokinetic mixers and extruders is that they operate in batch mode, instead of continuous processing, and do not require any external heaters. Heat to melt the thermoplastic polymer is supplied by frictional energy developed during high speed, high shear mixing. By processing in a batch mode, thermokinetic mixers can thoroughly mix to a degree unattainable in the continuous screw designs. Although some differences exist, the thermokinetic mixer operation begins with measured quantities of polymer and additives being fed through the top of the machine into a mixing chamber where blades of various designs and arrangements rotate at high speeds producing the desired modes of mixing, fluxing, and shearing. Once the material reaches a predetermined parameter (temperature, energy, or residence time) which occurs rapidly (15 to 30 seconds), it is discharged as a molten mass through a gate at the bottom of the machine. At this point, the process cycle is complete and the mixer can be charged with another batch. Material impingement against the blades and chamber wall thoroughly mixes and disperses the ingredients. Mixer operation variables that can be controlled are limited to blade or rotor speed, residence time, and indirectly, degree of heating/cooling. In tests sponsored by one manufacturer, the degree of mixing was found to be superior to both single- and twin-screw mixers (Kalb 1995a).

These high-intensity fluxing mixers operate with a powerful drive and are distinguished by a single rotor mounted with staggered blades placed at various angles. A typical thermokinetic mixer is shown in Figure 3.6. Revolving at high speeds (blade tip speeds up to 45 m/sec [148 ft/sec]), the rotor produces a fast-paced mixing action that causes a rapid temperature rise as intense shearing converts mechanical energy to heat. The combination of shorter production cycles and the lack of external heating helps reduce overall operational costs for a variety of thermoplastic processing applications. These mixers are also more flexible than extruders in processing blends of thermoplastic polymers, especially co-mingled recycled polymers. Other advantages include simplified operations, low maintenance costs, and high filler loading capacities (high waste loading capacity). High-intensity mixers can process a wide range and combination of ingredients, from pliable to extremely rigid materials, without changes to machine parts as might be required for screw extruders. Such versatility can reduce down time and simplify encapsulation processing requirements when several plastic/waste combinations are treated. Batch processing allows each particle to receive the same amount of work, improving control of residence time, temperature, and product uniformity, while simplified feeding increases the range of polymers and additives that can be compounded. High-shear mixers can process materials containing up to 50% (by weight) moisture, reducing or eliminating material pretreatment requirements. The extremely high blade speeds and rapid cycle times of high-shear mixers used because of batch charging and discharging can adversely affect productivity and batch-to-batch product consistency, as well as cause greater difficulty in precisely controlling melting and mixing.

Figure 3.6
Schematic of a Typical Thermokinetic Mixer

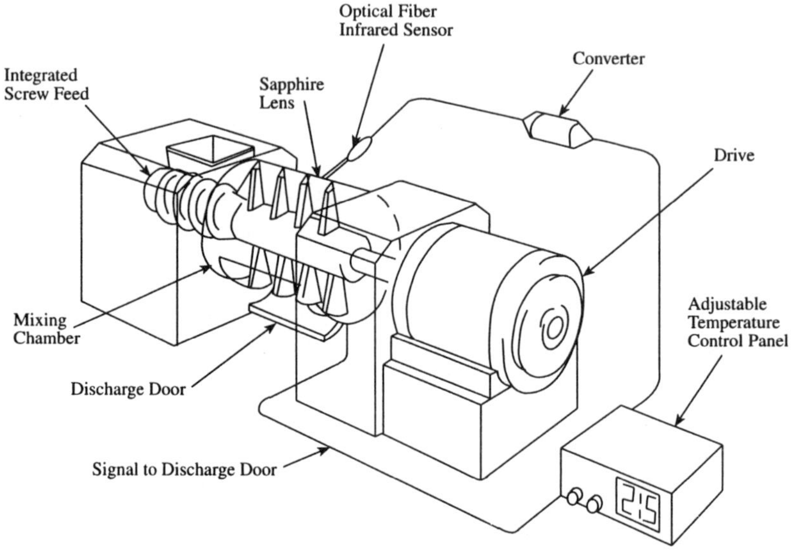

Source: Frados 1985

Currently, Brookhaven National Laboratory is working with EcoLex, Inc. (Jacksonville, Florida) to combine the advantages of thermokinetic mixing and single-screw extrusion in one cost-effective process. The thermokinetic mixer is used to remove residual moisture, reduce the size, premix, and melt the waste and thermoplastic polymer binder materials. The homogeneously-mixed, molten mixture is then fed to a single-screw extruder which continues to mix the materials and converts the process from batch to continuous operation. The combined process promises improvements in product performance and consistency, while providing increased processing flexibility for microencapsulation and macroencapsulation in one integrated system.

Other Processing Techniques. Other processing techniques for heating and mixing both polymer and waste are vertical, thin-film evaporators and screwless extruders. A thin-film evaporator is a drying device that takes advantage of an enlarged surface-to-volume ratio for quick and efficient heating and vaporization. Waste material and thermoplastic resins are fed at the top of the evaporator, spread into a thin layer, and conveyed through the device by gravity flow. Externally-applied heat simultaneously dewaters the waste and melts the polymer while material flow creates limited mixing for microencapsulation. The endproduct is a molten mixture that exits continuously out the bottom of the thin-film evaporator. This unique processing scheme takes advantage of waste pretreatment (drying) technology to melt and mix for polymer encapsulation. As a result, the need for waste pre-processing can be either reduced or eliminated. However, feed materials must be free-flowing; pretreatment in the form of grinding or reduction is required for large, bulky materials. Thin-film evaporators are also limited by high viscosity melts that occur with many thermoplastics and at high waste loadings resulting in disrupted gravity conveyance and material discharge problems. Despite using low molecular weight and low viscosity wax, tests conducted by Rocky Flats Plant indicate a limited maximum waste loading of 40% (by weight)(Faucette et al. 1994) compared to maximum waste loadings of 70% (by weight) using extrusion for selected wastes (Kalb, Heiser, and Colombo 1991a).

Screwless extrusion devices, including disk-type extruders, gear pumps, planetary-gear extruders, roller extruders, and ram extruders, have been in existence for years. Many of these devices were designed to meet unique requirements of specific processing applications and, therefore, lack the versatility and productivity found in conventional screw designs. To date, none have been used for waste encapsulation applications.

MACROENCAPSULATION. Macroencapsulation is advantageous when treating waste consisting of large or abrasive particles not suitable for processing through the extruder. The waste can contain large metal parts or fragments (e.g., contaminated equipment, drums, lead turnings), dry radioactive waste (e.g., trash, gloves, bottles), debris (e.g., decommissioning and demolition waste), or previously treated waste/waste form products (e.g., filters, degraded grout). The waste to be macroencapsulated is packaged in a porous structure or cage, placed in a drum with a slightly larger diameter, and the voids are filled with polymer. Numerous polymers, including polyethylene, epoxies or other thermosetting resins, and sulfur polymer, can be used to surround the waste and fill the voids. Single-screw extruders are ideal for polyethylene macroencapsulation applications, since they provide the most economical source of continuous molten plastic output. Mechanical rotation of the drum during processing prevents localized buildup of polymer and provides even filling of the annulus surrounding the waste. Figure 3.7 is a photograph of a bench-scale test sample that contains dry active waste macroencapsulated in polyethylene. The DOE has supported commercialization of polyethylene macroencapsulation, currently being implemented at Envirocare, Inc. (Salt Lake City, UT). Under DOE sponsorship, up to 227,000 kg (500,000 lb) of radioactively-contaminated lead is being treated by polyethylene macroencapsulation and disposed at Envirocare.

One disadvantage of using polyethylene for macroencapsulation is its inability to completely fill internal void spaces between the waste particles. Due to its high melt viscosity and the high thermal mass of the waste material to be macroencapsulated, polyethylene may cool and solidify before fully penetrating the waste. Although polyethylene macroencapsulation of lead is identified by the US EPA as a BDAT, the need for complete penetration of voids is not clearly defined from a regulatory perspective. Lower melt viscosity thermoplastics, such as sulfur polymer, might provide better penetration, but would still tend to solidify on contact with a large thermal mass. Preheating the waste would improve penetration of the polymer, but requires additional energy and an additional process step. Thermosetting resins, such as polyester styrene or epoxies, can be used to "flood" the container and better penetrate the voids. Vacuum or pressure can also be applied to enhance or accelerate penetration. A vacuum enhanced delivery system for thermosetting resins was demonstrated for macroencapsulating compacted dry active waste at Brookhaven National Laboratory (Franz, Heiser, and Colombo 1987). A gravity-fed delivery process for macroencapsulation using epoxy resins was developed by Rocky Flats Plant (Faucett et al. 1995).

UNIT SIZING. Unit sizing and design of processing equipment for polyethylene encapsulation of waste depends on:

- the type and volume of waste to be processed;

- operational considerations such as available space, staffing for operations and maintenance, need for remote handling, and storage capacity; and

- evaluation of capital and operating costs to provide an optimal balance in cost-effective operational efficiency.

Versatility to accommodate a broad range of wastes and intra-waste variability is an important design consideration. For example, significant

Figure 3.7
Photograph of a Typical Bench-Scale Sample of
Lead Wool Macroencapsulated in Polyethylene

Source: Kalb et al. 1995a

savings and process simplicity can be achieved if a single process is designed to handle both aqueous liquids and dry incinerator ash residues (microencapsulation), as well as miscellaneous debris and lead scraps (macroencapsulation). Extruders and mixing equipment described are commercially available in a wide range of sizes with potential output ratings ranging from several lb/hr to many ton/hr.

Space requirements can range from small bench-top units to large skids requiring several hundred square feet. The Brookhaven National Laboratory production-scale treatment system (rated at 900 kg/hr [2,000 lb/hr]) is configured as a fixed-base system and occupies approximately 28 m² (300 ft²). During the design phase, all components of a fully-integrated system must be considered. Capacities and physical space requirements for all ancillary equipment, such as pretreatment systems, materials handling systems, feeders, processing equipment, and monitoring and process control systems, must be compatible.

The need for remote handling and processing of waste materials provides an additional challenge for system designers. However, materials handling systems for waste and binder materials and remote drum handling equipment are readily available and have been used in the commercial nuclear power industry for many years. Process monitoring and control equipment can easily be installed away from processing areas to minimize operator exposure to radioactive and hazardous materials.

Polymer encapsulation processes are amenable to fixed-base or mobile skid-mounted configurations. For mobile units, separate components (e.g., pretreatment, extruder, material handling, process control) could be mounted in modules that are easily assembled and disassembled for shipping.

Unit sizing for macroencapsulation depends on the size of debris being treated and the volume of waste. As discussed previously, extruders are readily available in numerous sizes with output rates ranging from 9 kg/hr (20 lb/hr) to several tons/hr. Prototypical 208 L (55 gal) drums of debris have been successfully processed to date. A typical production-scale 11.4 cm (4.5 in.) single-screw extruder, rated at 900 kg/hr (2,000 lb/hr), can process a 208 L drum of macroencapsulated waste in about 18 minutes. Envirocare has installed a 11.4 cm single-screw extruder, as described, at their facility in Clive, Utah, to process DOE mixed waste lead and debris.

3.2.3.2 Sulfur Polymer Cement Encapsulation

A major advantage of SPC is the simple design required for stabilizing waste. A process-flow diagram for SPC is shown in Figure 3.8. A discussion of the equipment required and its operation is provided below.

Figure 3.8
Process-Flow Diagram for Sulfur Polymer Cement

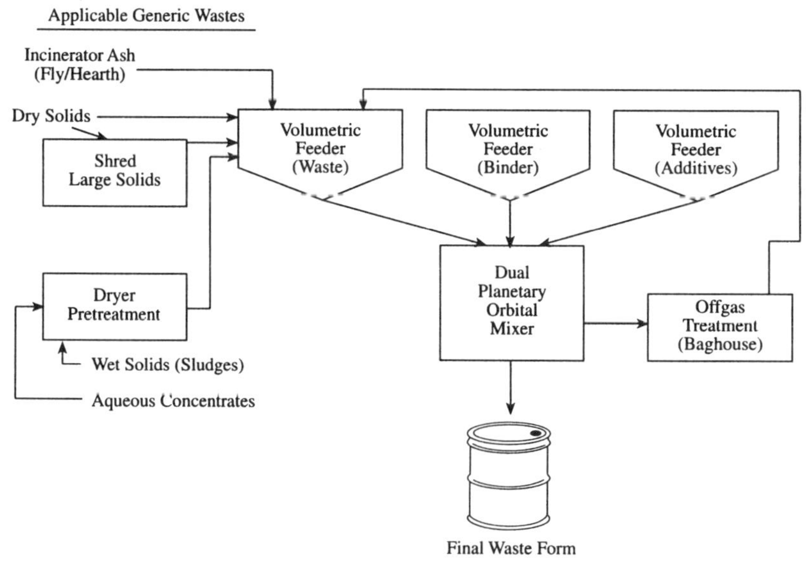

Feeding the waste and SPC into the mixer must be done precisely in order to ensure proper waste loading. Brookhaven National Laboratory has shown that either volumetric or loss-in-weight feeding (Kalb 1995b) effectively control material flow rates. The same type of feeders can be used to feed additives, if necessary.

Bench-scale development work at Brookhaven National Laboratory was conducted using several types of heated batch mixers including low- and high-shear blade mixers and double planetary orbital mixers. The latter provides a highly-efficient mixing pattern at relatively low mixing speeds, thus reducing air and gas entrainment in the molten mixture. Heating can be achieved by thermocouple-controlled electric resistance band heaters, steam, or hot oil circulation heaters (Kalb and Colombo 1985a; Kalb, Heiser, and Colombo 1991b; Adams and Kalb 1996).

Two types of production-scale mixers have been evaluated by Idaho National Engineering Laboratory (Darnell 1993). One is a Holo-Flite mixer produced by Denver Equipment Company. The other is a Porcupine Processor. Both mixers use hot oil as the heating agent. They pump the oil through the hollow shafts, flights, and external jacket. Both can be easily modified to extend the heated jackets to heat the shafts. This will allow the re-melting of solidified waste product without damaging the mixer (Darnell 1993).

The Holo-Flite mixer is shown in Figure 3.9 and has a dual shaft with 18 cm (7 in.) diameter flights. Each of the shafts is approximately 4.0 m (13 ft) long. This horizontal mixer can place heat within 5.1 cm (2 in.) of every waste particle, thus preventing hot spots and allowing remelt after shutdown. As a closed system, the mixer is designed to confine dust and gases by operating at a negative differential pressure (Darnell 1993). The Porcupine Processor, as shown in Figure 3.10, has a single shaft with 30 cm (12 in.) flights. The shaft is approximately 1.8 m (6 ft) long. Instead of having a continuous screw, this processor has hollow paddles shaped like beaver tails (Figure 3.11). Scientific Ecology Group, Oak Ridge, Tennessee, has conducted pilot-scale testing of SPC encapsulation for mixed waste fly ash using a steam-heated, high-shear mixer. They have reported successful results for processing batches of approximately 1,360 kg (1.5 ton).

Some mixers allow processing under negative pressure conditions. Operating under vacuum reduces the necessary processing temperature and facilitates capture of any offgases that are generated.

3.2.3.3 In Situ Polymer Stabilization/Solidification

A number of technologies that have been used to place conventional grout walls and curtains can be used with in situ polymer S/S. These include deep soil mixing, jet grouting, and permeation grouting. Other methods, such as excavation or trenching and displacement/replacement techniques, involve

Figure 3.9
Holo-Flite Mixer

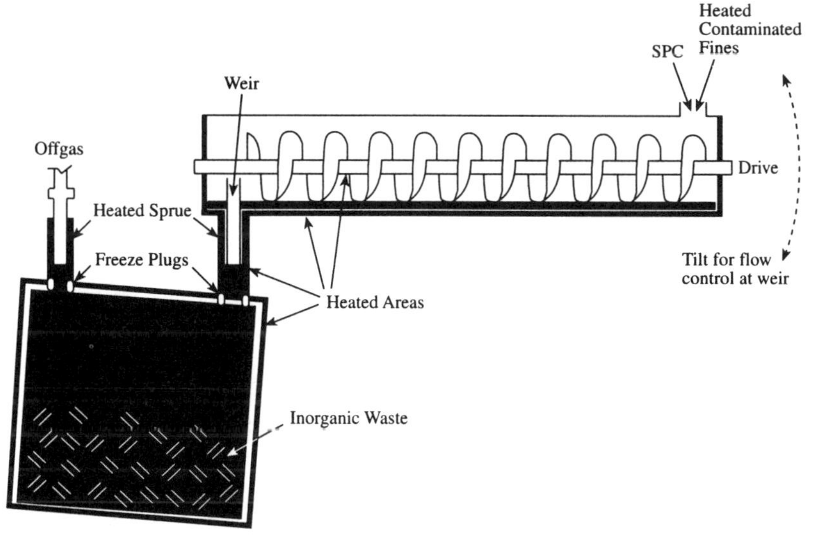

Source: Darnell 1993

removal of the soil, treatment, and replacement in the ground. Strictly speaking, such technologies are ex-situ. Deep soil mixing relies on a vertical auger to bore a large diameter hole in the soil and mix the soil with the polymer, which is injected under low pressure (~2.1 MPa [~300 psi]). Jet grouting uses a rotating pipe that is drilled into the soil. The polymer is injected through a small orifice (1 mm [39 mil] diameter) at high pressure (~34 MPa [~5,000 psi]) as the pipe is gradually withdrawn from the soil. Penetration is typically 1.5 to 2.0 m (4.9 to 6.6 ft) in diameter. A recent modification to jet grouting, known as "super jet-grouting" uses very high pressures (up to 170 MPa [25,000 psi]) to hydrofracture and inject the grout. This technique might be useful in treating sites where buried waste in the form of drums and containers needs to be breached for complete treatment. Permeation grouting injects low viscosity materials through a series of injection wells at low pressure (~0.69 MPa to 1.38 MPa [~100 to

Figure 3.10
Porcupine Processer

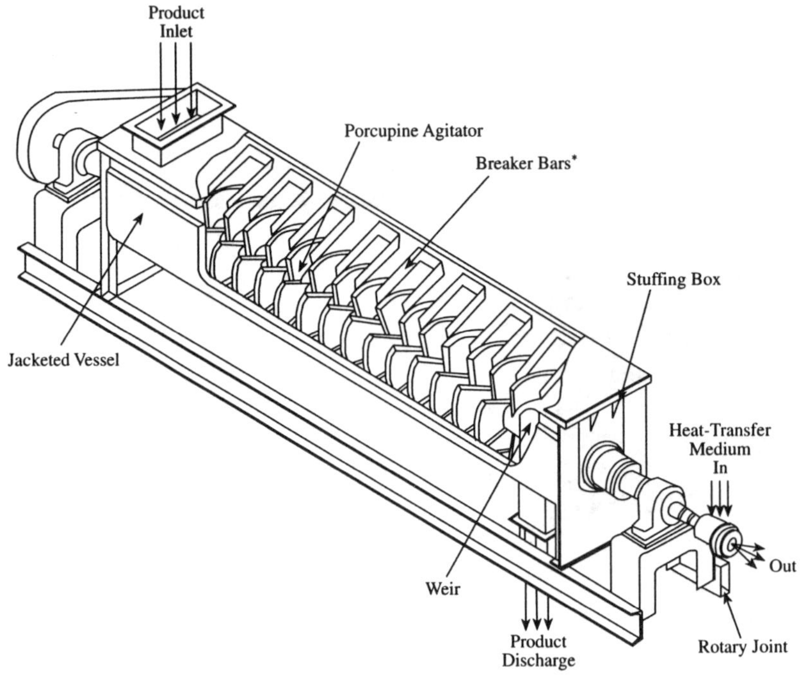

Product
Inlet

Porcupine Agitator

Breaker Bars[*]

Stuffing Box

Jacketed Vessel

Heat-Transfer
Medium
In

Out

Weir

Product
Discharge

Rotary Joint

[*]Opposing breaker bars, on opposite wall of vessel, not shown.

Source: Darnell 1993

200 psi]) and relies on permeation through the soil for penetration. A recent evaluation of these methods for use with emplacement of barrier walls (McLaughlin et al. 1992) recommended permeation grouting and jet grouting as the most effective methods for placing polymer barriers. These techniques are illustrated in Figures 3.12 and 3.13. Permeation grouting is susceptible to hydrofracturing and short-circuiting which make it difficult to control the flow of grout. Deep soil mixing is effective for contaminated soils, but might not be applicable for buried waste; penetration of the augers might be hindered by the waste.

Figure 3.11
Porcupine Processer Paddles

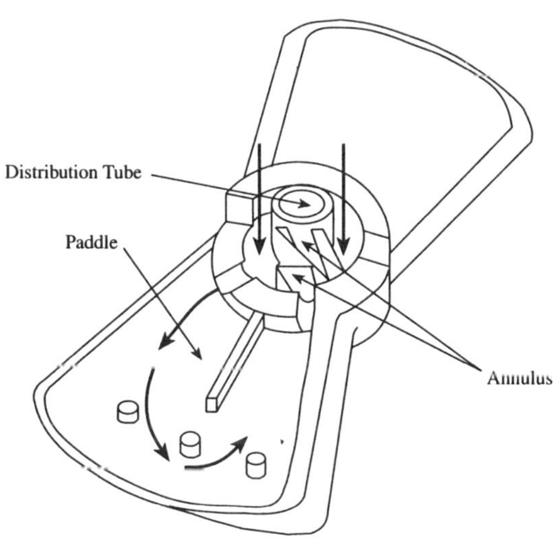

Source: Darnell 1993

Figure 3.12
Subsurface Barrier Installation by Permeation Grouting

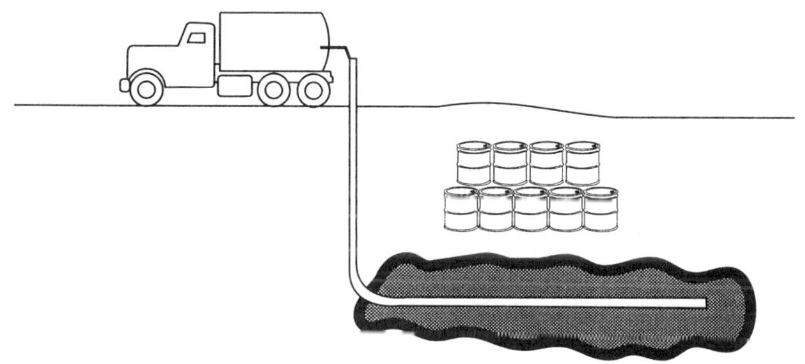

Source: Heiser and Milian 1994

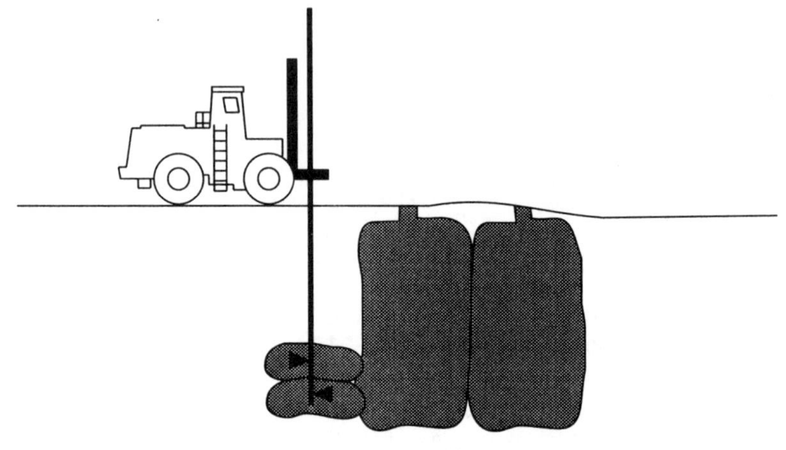

Figure 3.13
Conventional Column Jet Grouting

To select the most appropriate in situ technology, the following factors must be considered: the placement effectiveness, site soil conditions, presence of buried waste (other than soil), volume of soil to be treated, and economic feasibility. It is also important to remember that treatment of waste contaminated soil is different than placing a barrier when extrapolating barrier placement data.

3.2.4 Process Modifications

3.2.4.1 Polyethylene Encapsulation

A major advantage of polyethylene S/S is that it does not require a chemical reaction to solidify and therefore, is not usually affected by changes in waste chemistry over time. Varying the type of waste to be processed or the physical characteristics of a given waste stream might require modifications in pretreatment requirements, waste loading specifications, polymer type, feed mechanism, feed rates, process rates, or zone temperatures. When variations in waste types and/or conditions are anticipated, the system should be designed for maximum flexibility. In some cases, modifications in the extruder screw design might provide optimal processing results. For example, one or more alternative screw designs should be kept on hand in case they are needed. Many varieties of polymers with a wide range of melt temperatures, melt viscosities, and mechanical properties are available both as virgin materials and as recycled feedstock. Varying the type(s) of polymers or the mix ratio of several feedstock polymers can dramatically affect processability and final product performance. While the extrusion process is generally forgiving, varying waste loading specifications, feed rates, and zone temperatures is usually required to optimally process a given waste stream. Additives might be incorporated to reduce leachability of contaminants. Several materials such as sodium sulfide and sodium hydroxide have been used to reduce solubility of toxic metals and hence reduce leachability of the encapsulated final waste forms.

3.2.4.2 Sulfur Polymer Cement Encapsulation

Depending on the type of waste and its chemical and physical characteristics, process modifications might be required. For example, additives to enhance mechanical integrity (e.g., glass fibers) can successfully mitigate the expansive forces of soluble salts that can cause cracking when the waste form is exposed to saturated conditions. Brookhaven National Laboratory added 0.5% (by weight) of glass fibers into SPC for ash wastes that were high in chloride salts to eliminate cracking attributable to immersion in water (Kalb, Heiser, and Colombo 1991b). High concentrations of toxic heavy metals can cause the encapsulated waste to fail US EPA TCLP concentration limits. Additives have also been shown to decrease leachability of toxic metals when excessive concentrations of metals are present in the waste. Brookhaven National Laboratory has reported that the addition of sodium sulfide when mixed with the waste and SPC can effectively stabilize the toxic metals. The sodium sulfide reacts with metal salts to form a low solubility metal sulfide within the microencapsulated waste form. A ratio of 0.175:1, sodium sulfide to fly ash was reported as an effective ratio to enable the treated ash waste (which contained 7% [by weight] soluble lead salts) to pass the TCLP (Kalb et al. 1991).

3.2.4.3 In Situ Polymer Stabilization/Solidification

In situ applications are, by definition, more diverse than ex-situ. Therefore, site-specific conditions including soil characteristics (composition,

particle size, permeability, moisture content), volume, and remediation requirements will foster a need for process modifications from one application
to another. In some cases, the type of polymer will need to be varied based
on cost, viscosity, chemical compatibility, etc. In other cases, the placement
technology will require tailoring to the specific conditions of the site. For
example, if the site contains a high concentration of low permeability clays,
a placement technique that provides more active mixing (e.g., deep soil au-
guring) rather than one that relies on permeation through the soil, would be
preferred. Once a given technique and material are selected, modifications
can still be made to improve system performance. For example, quantities
of catalyst and promoter can be varied depending on requirements of the
soil. Lower concentrations of catalyst and promoter may decrease peak
exothermic temperatures (reducing volatility), but will also decelerate setting
and curing times. Grouting pressures and drilling rates can also be adjusted
to compensate for varying site conditions. Higher pressures generally result
in increased penetration and may provide improved homogeneity of the
waste-binder mixture.

3.2.5 Pretreatment Processes

3.2.5.1 Polyethylene Encapsulation

To be efficiently processed by polyethylene extrusion, waste materials
must be dry and meet acceptable particle size specifications. If as-generated
waste does not meet these processing requirements, pretreatment processes,
including drying and size reduction, is needed. Waste to be encapsulated in
polyethylene should not contain more than 1 to 2% (by weight) moisture.
Various commercial drying mechanisms including spray dryers, fluidized
bed dryers, thin-film evaporators, and vacuum dryers, have been investigated
for use with polyethylene-based systems. Vacuum drying was selected at
Brookhaven National Laboratory because of its ability to meet both moisture
and minimum particle size requirements. The commercially available
RVR-200 vacuum dryer, supplied by MMT of Tennessee, consists of a
horizontally-mounted stirred mixer, steam generator, condensate recovery
skid, and chiller (Figure 3.14). It can dry up to 760 L/day (200 gal/day), but
larger systems are available with outputs up to 3,030 L/day (800 gal/day).
The dryer is controlled and monitored remotely and includes closed circuit
video for observing the condition of the mixture.

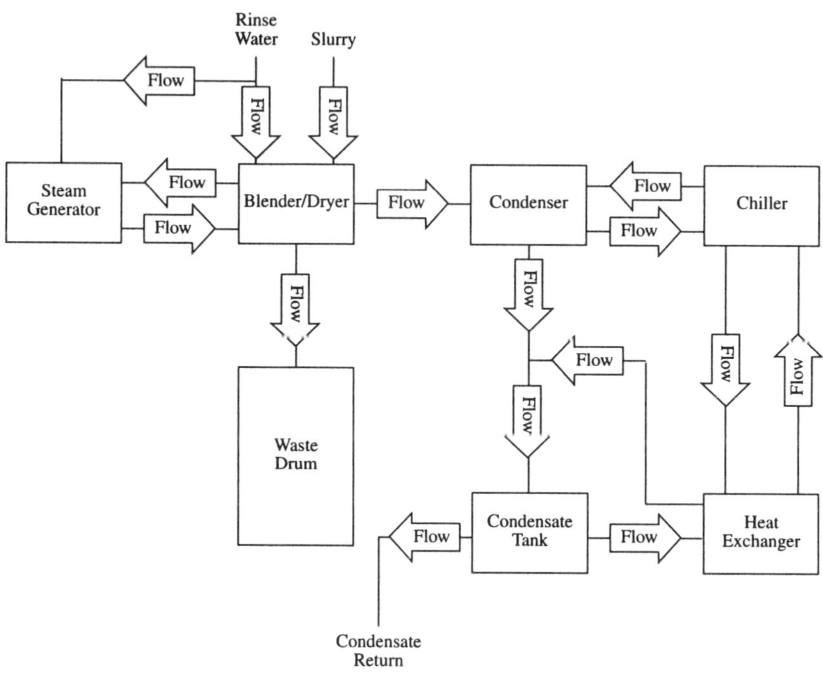

Figure 3.14
Flow Diagram for RVR-200 Vacuum Dryer

Particle size requirements vary according to the type of processor selected. In general, the smaller the particle size, the more effective the microencapsulation of waste particles, and the lower the leachability of the final waste form. Therefore, for particles >3 mm (0.12 in.), performance is improved by size reduction prior to extrusion. The Brookhaven National Laboratory pretreatment system includes an in-line comminutator at the dryer discharge. It consists of a series of rotating knife blades that break up particles and force them through a mesh screen. A typical resulting particle size distribution is shown in Figure 3.15.

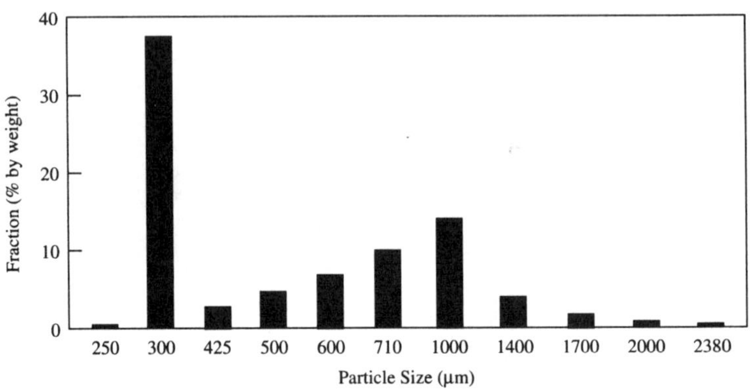

Figure 3.15
Particle-Size Distribution of Nitrate Salt Surrogate Following Pretreatment

Thermokinetic mixers are currently being tested to pretreat wastes with low- to moderate- moisture content. The advantages of this approach are lower energy requirements for drying, lower capital costs, and enhanced homogeneity of waste and binder. However, wastes with high moisture content reduce available frictional heat energy and inhibit the mixer's ability to drive off moisture. These wastes require more conventional dryer pretreatment.

3.2.5.2 Sulfur Polymer Cement Encapsulation

As with polyethylene encapsulation, smaller waste particles allow more effective microencapsulation of the waste and minimize leachability. Pretreatment size reduction with a shredder or a crusher may be required to achieve this small particle size. Unlike LDPE, however, SPC has no limit on minimum particle size, since it is a low viscosity material and is processed by batch mixing.

Sulfur polymer cement is also incompatible with water in the waste. Moisture that is volatilized during processing can become entrained, lowering the product density and mechanical integrity, and increasing porosity and leachability. Therefore, all waste should be dried prior to stabilization (<1% moisture). If wastes are heated and held at the process temperature prior to mixing with sulfur polymer, residual moisture is liberated, and the waste/binder mixture remains in a flowable state for a longer period of time (Mayberry et al. 1993). Since SPC is compatible with a wider range of acceptable particle sizes than is LDPE, the choice of dryer for pretreatment is less critical.

3.2.6 Posttreatment Processes

Since polymer encapsulation processes operate at relatively low temperatures, most contaminants (e.g., radionuclides, such as Cs, Sr, Co, and toxic metals, such as Pb, Cr, Cd) are not volatilized during processing, minimizing offgas posttreatment requirements. However, highly volatile metals, such as Hg, as well as sound health and safety engineering practices, require the use of offgas collection and ventilation systems for extruders, dryers, or other process equipment that generate gaseous or particulate effluents. For streams containing small quantities of volatile organics, carbon traps can be used to absorb organic vapors prior to discharge.

The final waste form geometry can be determined to meet the needs of the particular application. For example, final waste form products can be produced as standard cylindrical or cubical monoliths that require less room for storage. Leaching of the microencapsulated final waste form can be decreased further by surrounding the waste form with a clean layer of plastic (double encapsulated). This can accomplished by co-extrusion or a plastic liner. Alternatively, encapsulated wastes can be pelletized and stored in secondary containers or encapsulated within a larger waste form.

3.2.7 Process Instrumentation and Controls

3.2.7.1 Polyethylene Encapsulation

Ames Laboratory has developed an on-line monitoring system for the polyethylene encapsulation process (Wright et al. 1994). Based on Transient Infrared Spectroscopy (TIRS), the system enhances quality assurance/quality control (QA/QC) of the process by assuring that the preset mixture ratio is being delivered and preserved throughout the process. The Ames monitor is shown in Figure 3.16. The computerized monitor compares real-time process spectra with previously stored calibration data to produce instantaneous and time-averaged waste loading data. Calibration data showing predicted waste loading using the TIRS monitor plotted against actual waste loading (Figure 3.17) demonstrate the accuracy of the monitor. Information is also stored for a permanent QA record of the processing run.

Figure 3.16
Transient Infrared Spectroscopy (TIRS) On-Line Monitoring System Developed by Ames Laboratory for the Polyethylene Encapsulation Process

The full-scale polyethylene encapsulation facility at Brookhaven National Laboratory includes instrumentation and process monitoring modules that can be remotely located from the processing equipment to reduce operator exposure to radiation and hazardous materials. The extruder module includes screw speed control, solid state monitoring, and control of heating and cooling in five separate barrel and two die zones, and digital and analog process indicators for melt temperature, melt pressure, screw speed, and current draw. The integrated process control unit is operated on a personal computer using a customized program written for Windows™-based control software. The controller receives a weight signal from a solid state drum scale, integrates the information into a process rate and operates the feeder master control unit to keep pace with the process. Thus, as the extruder screw speed is increased, the output increases and the feeder rates are increased proportionally to maintain the process rate (Kalb et al. 1995b).

Figure 3.17
TIRS Monitor Plotted vs. Actual Waste Loading

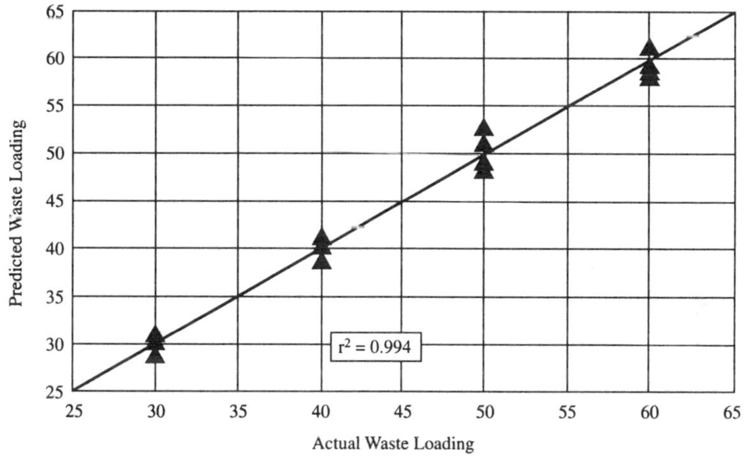

Source: Kalb and Lageraaen 1994

3.2.7.2 Sulfur Polymer Cement Encapsulation

Sulfur polymer is processed by stirred batch heaters, which are relatively simple mixing systems. Parameters to be monitored are melt temperature, system vacuum, and mixer speed. A closed circuit video monitor allows the operator to observe the condition of the mixture.

3.2.7.3 In Situ Polymer Stabilization/Solidification

Combining the catalyst and promoter with the monomer is best accomplished by mixing the proper ratio of each component with half the monomer and then pumping the two mixtures simultaneously through twin grout nozzles. Flow rates of each mixture must, therefore, be carefully controlled and monitored to assure that the proper ratios of components are delivered and mixed with the waste. Grouting pressure controls the extent of radial penetration, and drill speed regulates how quickly the column can be placed. Each of these parameters requires operator attention to ensure proper installation of the in situ polymer column.

3.2.8 Safety Issues

3.2.8.1 Polyethylene Encapsulation

Worker exposure to radioactive and hazardous materials during processing and maintenance operations is an important safety issue for any treatment system. Because of relatively simple and reliable operation, polymer processing equipment generally requires less maintenance than other types of S/S systems. For example, extruders can be readily purged and cleaned by processing polymer that contains no waste or by using specially-developed, purging compounds. In the event of an unplanned shutdown, polymer solidified within the extruder can be easily re-melted enabling the operator to rapidly resume processing. Remote handling, monitoring, and process controls also minimize exposure to hazards. Standard OSHA safety practices for operating equipment at elevated temperatures (e.g., heat shields, protective clothing) ensure safe operation of polymer processing equipment.

3.2.8.2 Sulfur Polymer Cement Encapsulation

Airborne SPC dust can be mildly explosive so safety precautions must be exercised. The U.S. Department of Transportation does not consider SPC to be flammable. When exposed to a direct flame, SPC will burn, but will self extinguish when the flame is removed (Mayberry et al. 1993). SPC will emit hydrogen sulfide gas, sulfur dioxide gas, and volatile organic sulfur compounds if SPC is melted above 160°C (320°F). These gases are both poisonous and flammable (Mattus and Mattus 1994); therefore, precautions must be observed to avoid overheating SPC.

The flash point of SPC as determined by the Cleveland Open Cup method is 177°C (351°F), and the auto-ignition temperature is 232 to 254°C (450 to 489°F).

3.2.8.3 In Situ Polymer Stabilization/Solidification

Polymerization reactions are exothermic and, depending on the types and quantities of catalyst and promoter used, can be difficult to control. However, if monomers are stored without added catalyst or promoter, and the combining of the catalyzed and promoted monomers takes place at the time of placement in the ground, few difficulties have been experienced. Thermal energy can also act in a similar manner as chemical promoters, so materials should be stored in a cool area prior to use.

3.2.9 Specification Development

3.2.9.1 Polyethylene Encapsulation

In addition to the typical engineering requirements, specification should take into account materials of construction, power requirements, and the integration of ancillary components. Many waste streams are either acidic or alkaline and can be highly corrosive. Process components that come in direct contact with corrosive materials should be of appropriate corrosion-resistant alloys. Specification of each of the required auxiliary components should be coordinated with the overall system requirements. For example, pretreatment equipment must complement final waste form processing in terms of production rates, product specifications (e.g., moisture content, particle size), and physical layout. Handling systems must be able to supply materials in either batch or continuous mode, as appropriate, at required production rates. All equipment must meet appropriate engineering standards for safety. Vendors with experience in producing equipment for nuclear and hazardous materials should be able to demonstrate compliance with appropriate certification standards (e.g., American National Standards Institute [ANSI]). Thermoplastic polymers are available in a wide range of densities, melt temperatures, melt viscosities, and as either virgin or recycled feedstock. Therefore, the type of polymer should be selected according to the type of waste being treated and feedstock specifications written into the waste treatment plan and operating procedures.

3.2.9.2 Sulfur Polymer Cement Encapsulation

In choosing an appropriate mixing system, the engineer must consider which method of thermal input will be used (e.g., steam, hot water, oil, or electric resistance heating) and which type of mixing blade design is optimal. The latter choice should consider the nature of the waste stream that is anticipated. For dry powders with small particles, slow-speed orbital or paddle mixers might be ideal. For wastes with larger particle size and/or greater density, high shear mixing might be preferable. Prior to selecting the mixer design, bench-scale vendor testing to confirm applicability is often required. Offgas treatment should include a carbon trap to reduce noxious (but harmless) odors that are associated with molten sulfur. To expedite batch processing, the SPC encapsulation process can be designed with a secondary heating tank to pre-melt the binder and introduce it in liquid form into the mixing vessel. Likewise, if the waste is preheated prior to charging into the mixer, overall process time is reduced and mixing can be improved. Engineering design specification requirements are similar to other polymer encapsulation technologies, such as polyethylene, in that the system must be able to withstand corrosive conditions and meet industry certification standards.

3.2.9.3 In Situ Polymer Stabilization/Solidification

The key factors in selecting monomers, catalysts, promoters, and additives, if required, are the type and properties of the soil being remediated (e.g., particle size, permeability, chemical composition) and the contaminants of concern. Key engineering properties of the polymer matrix, such as viscosity, setting time, and peak exotherms, can be tailored according to the specific soil and contaminant conditions present. For example, the choice of catalyst and promoter and specification of mixing ratios can affect how quickly polymerization occurs. Setting and curing times must be tailored to ensure rapid solidification without interfering with on-going drilling and emplacement operations. Polymer/soil mixing ratios are predetermined from laboratory-scale testing to optimize performance of the solidified product (e.g., leaching), while providing cost-effective treatment (e.g., minimizing the quantity of polymers added). Design specifications for emplacement equipment must consider issues of pumping and delivery pressure, nozzle design, penetration, maintenance/cleaning, etc.

3.2.10 Cost Data

3.2.10.1 Polyethylene Encapsulation

In general, capital and operating costs for polyethylene encapsulation systems are moderate to higher than those of conventional hydraulic cement grout processes but lower than high temperature vitrification systems. The cost of polyethylene feedstock varies between $0.99 to $1.32/kg ($0.45 to $0.60/lb) depending on the type of polymer and the quantity purchased. Use of recycled polymer feedstock might reduce material costs in the future, but the availability of well-characterized recycled polymers is currently limited. Material binder costs can be estimated considering that microencapsulation waste loadings typically range from 50-70% (by weight). Typical capital costs and estimated energy requirements for production-scale process equipment are summarized in Table 3.12. Several equipment options are presented for comparison. It should be noted, however, that the total cost of treatment is often influenced most by storage, transportation, and disposal costs, which, in turn, are directly related to final waste form volumes produced.

Table 3.12
Typical Capital Cost Data and Estimated Energy Requirements for Production-Scale Polyethylene Encapsulation Process Equipment[a]

Component	Estimated Capital Cost ($)	Estimated Energy Requirements (kW)
Single Screw Extruder	200,000	50
Twin Screw Extruder	450,000	100
Thermokinetic Mixer	130,000	40
Vacuum Dryer	550,000	7[b]
Material Handling System	30,000	c
Feeder System	35,000	c
Process Control System	15,000	c

[a]Based on unit sizing with approximate total output of 2,000 lb/hr
[b]Per pound of salt produced
[c] Negligible

3.2.10.2 Sulfur Polymer Cement Encapsulation

Since processing is carried out using simple stirred melters, capital and operating costs for sulfur polymer encapsulation are slightly lower than for higher viscosity thermosetting polymers. Production-scale mixers can cost between $100,000 to $150,000, while pretreatment and feeder equipment are similar in cost to those described in Table 3.12. The cost of the SPC raw material is currently about $0.26/kg ($0.12/lb), slightly higher than conventional hydraulic cements. SPC binder costs can be estimated based on typical waste loadings ranging from 40 to 60% (by weight). However, the cost of the material is just one factor in determining the life cycle cost of treatment and disposal. All aspects of the cost formula, such as pretreatment requirements, energy, operation and maintenance, and waste form storage/transportation/disposal costs must be considered. Because SPC encapsulation can, in some cases, reduce final waste form volumes by a factor of up to 3 times as compared with Portland cement concrete, SPC should prove cost-effective compared with conventional hydraulic cement technologies.

3.2.10.3 In Situ Polymer Stabilization/Solidification

Capital costs for in situ polymer S/S processing equipment range from around $550,000 to $2,000,000 for a typical jet grouting rig. Approximately 200 kW of energy is required to operate drilling and high pressure pumping equipment. Material costs for thermosetting polymers are relatively high and can range from around $1.74/kg ($0.79/lb) for polystyrene and some acrylics to more than $14.33/kg ($6.50/lb) for polysiloxanes and some epoxies.

3.2.11 Design Validation

As with other S/S technologies, the appropriateness of specific engineering system design should be evaluated for each new polymer S/S application. Typically, this is done through pilot- or full-scale testing using surrogate waste materials. Testing can be accomplished by technology developers, but is most effective if conducted by or in conjunction with commercial vendors that plan to market the technology. If competing designs or processes are being considered for a particular remediation project, treatability and pilot-scale testing of each system is recommended to provide sufficient data to make an informed evaluation. Peer review of polymer S/S technologies have been conducted by tapping expertise within DOE, the regulatory community (US EPA and NRC), academia, and the commercial sector.

3.2.12 Permitting Requirements

Treatability studies for characteristic hazardous and mixed wastes are required to demonstrate compliance with RCRA by 40 CFR 261 as administered under the US EPA or authorized state agencies with jurisdiction over environmental issues. Treatability studies must include the TCLP test. US EPA recently proposed significantly reducing the allowable concentrations of toxic metals in TCLP leaching (US EPA 1995b). Small-scale treatability studies (treatment of <10,000 kg/yr [22,000 lb/yr]) can be conducted without special permits as long as the appropriate authority (state agency or US EPA) is notified at least 45 days prior to initiating the study. For larger quantities, a RCRA Research, Development, and Demonstration (RD&D) treatment permit is required. For some listed waste streams, US EPA has established technology-based standards, eliminating the requirement for treatability studies.

For commercial radioactive or mixed wastes, treatment must meet performance based standards defined in 10 CFR 61 by the NRC (NRC 1983). The NRC requires the waste generator or treatment vendor to prepare a topical report that documents results of specific final waste form performance tests for stability and leaching characteristics under a variety of simulated disposal scenarios (NRC 1991a). The stability and performance tests include compressive strength, water immersion, thermal cycling, biodegradation, radiation stability, and leachability. Topical reports submitted to NRC are reviewed, and if deemed acceptable, the technology is licensed for commercial treatment.

The DOE's disposal requirement is based on a risk-based performance assessment analysis for a specific disposal site (DOE Order 5820.2A). Under current law, DOE must comply with disposal requirements established by US EPA, but is exempt from NRC requirements. However, some DOE sites are using NRC's test protocol because DOE has not defined specific testing requirements to support DOE Order 5820.2A (Mayberry et al. 1993). Individual disposal sites or states can impose their own waste acceptance criteria (WAC) for DOE low-level radioactive wastes. Currently, most site

WACs are less stringent than NRC performance criteria. For example, typically they require no free water or they impose maximum radioactive concentrations. As states and regional state compacts (groups of states agreeing to cooperate in the siting of new disposal facilities) continue to construct new facilities, more rigorous standards might be adopted.

Regulatory issues associated with the use of polymers for in situ barrier materials have been investigated (Siskind and Heiser 1993). They concluded that since polymers have been used extensively in the construction industry in full compliance with applicable health, safety, and environmental regulations, no difficulties are anticipated for their use as in situ barriers. The same set of regulations, with the possible exception of the Safe Drinking Water Act, should pertain to polymer injection used in road construction, subsurface barriers, or S/S of contaminated soils. Of course, the performance of the remediated site must be in accord with all pertinent environmental regulations.

3.2.13 Performance Measures

Performance is measured in terms of both processing efficiency and the final waste form's ability to retain contaminants over time. As with other S/S technologies, processing efficiencies are expressed as the quantity of waste that can be effectively encapsulated per unit volume, while still maintaining adequate performance, i.e., ability to meet regulatory and disposal site acceptance criteria. In order to make comparisons on a equivalent basis, these data are usually presented in terms of dry weight percent of waste. Polyethylene encapsulation has been successfully demonstrated at waste loadings from 30 to 70% (by weight) dry waste, depending on the type of waste, levels of contaminants, and the performance standards required.

Mechanical integrity, durability, and leaching characteristics are the critical performance measures of potential final waste form behavior under long-term storage and disposal conditions. Low-level radioactive and mixed wastes generated in the commercial sector are subject to NRC licensing requirements for treatment and disposal described in Section 3.2.12. The NRC has established minimum waste form performance requirements for durability and leaching in support of 10 CFR 61.

Typical performance test data for polyethylene encapsulated waste forms are presented in Tables 3.13 and 3.14. The results of compression strength tests for SPC are listed in Table 3.15. Sulfur polymer cement provided a

Table 3.13

Typical Durability and Leaching Data for Polyethylene
Microencapsulated Final Waste Forms Containing 60%
Simulated Nitrate Salt Waste by Weight

Final Waste Form Performance	Test Protocol	Results	Minimum Standards for NRC Licensing
Compressive Strength	ASTM D–695	2,200 psi	60 psi
Water Immersion	90 day; ASTM –695	2,310 psi	60 psi
Thermal Stability	30 cycles, – 40+60°C, ASTM D–695	1,930 psi	60 psi
Biodegradation	ASTM G21, G22; ASTM D–695	1,460 psi[*]	60 psi
Radiation Stability	10^8 rad; ASTM D– 695	2,420 psi	60 psi
Radionuclide Leachability	ANS 16.1	9	Leach Index $\geq$ 6.0

[*]Apparent loss in strength following biodegradation was attributed to test protocol rather than structural properties of the waste forms.

Adapted from Kalb, Heiser, and Colombo 1991a; Franz, Heiser , and Colombo 1987

Table 3.14

Typical ANS 16.1 Leach Test Data as a Function of
Waste Loading for Microencapsulated Final Waste
Forms Containing Simulated Nitrate Salt Waste

Waste Loading (% by weight)	Cumulative Fraction Leached	Leachability Index[*]
30	0.9	11.1
50	6.3	9.7
60	15	9
70	73.4	7.8

[*]Conducted as per procedures outlined in ANS 16.1 Standard Leach Test Method. Minimum Leach Index recommended by NRC is 6.0.

Adapted from Kalb, Heiser, and Colombo 1991a

Table 3.15
Compressive Strength Data for Sulfur Polymer/Ash
Waste Forms Following NRC Performance Testing

Test Protocol	Compressive Strength[a,b]
Compressive Strength	4,250 psi
Water Immersion	3,870 psi
Thermal Cycling	3,870 psi
Biodegradation	2,620 psi
Radiation Stability	1,950 psi

[a]Data for waste forms containing 30% ash by weight. Biodegradation and Radiation Stability testing conducted on neat sulfur polymer specimens (no waste).
[b]Minimum compressive strength recommended by NRC is 500 psi.

Source: Kalb, Heiser, and Colombo 1991b

Table 3.16
ANS 16.1 Leach Data for Sulfur Polymer Final
Waste Forms Containing Incinerator Ash

Ash Waste Loading (% by weight)	Co-60 Leachability Index[*]	Cs-137 Leachability Index[*]
20	14	11.2
40	14.6	11.1

[*]Conducted as per procedures outlined in ANS 16.1 Standard Leach Test Method. Minimum Leach Index recommended by NRC is 6.0.

Source: Kalb and Colombo 1984

compression strength performance of at least four times the minimum NRC requirements of 3.4 MPa (500 psi) on all tests. Table 3.16 illustrates the leach performance data for the ANS 16.1 leach test recommended by NRC (ANSI/ANS 1986; NRC 1991a). The SPC leachability indices are four to eight orders of magnitude lower than those required by NRC. The performance and durability of in situ polymer concrete solidification for treatment of buried wastes at Idaho National Engineering Laboratory are currently under investigation. Preliminary data for in situ stabilized soils shown in Figure 3.18, indicate excellent integrity and durability of treated surrogate soil waste samples.

Figure 3.18

Compressive Strength of Polymer Soil Grouts After Resistance Testing

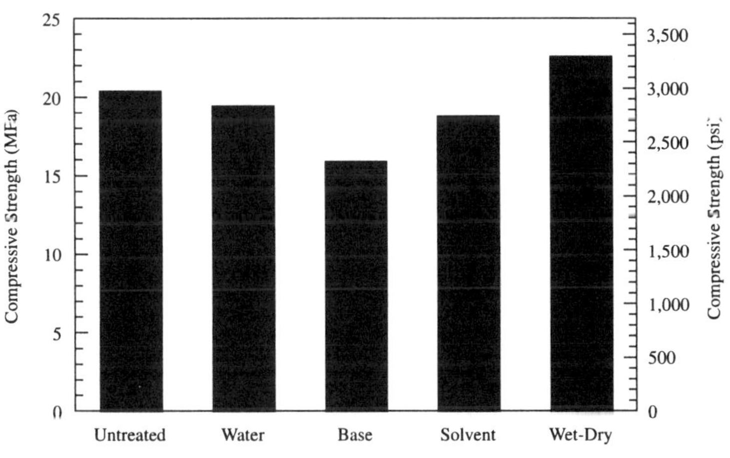

Source: Heiser 1995

For hazardous and mixed wastes, US EPA's TCLP criteria apply. The test was designed for a traditional hydraulic cement S/S process and requires size reduction of the monolith to pieces that can pass through a 9.5 mm (370 mil) sieve. Furthermore, the test is biased toward alkaline pH-based systems, since solubility of most metals is limited under high pH conditions. Given size reduction requirements and leachate pH conditions, processes that rely primarily on microencapsulation are at a distinct disadvantage by comparison. Nevertheless, all potential S/S technologies must meet 40 CFR 261 TCLP leaching protocol. Typical TCLP leaching data for polyethylene encapsulated waste, compared with untreated baseline data, are presented in

Table 3.17
Typical Toxicity Characteristic Leaching Procedure Data for Polyethylene Microencapsulated Waste

Waste Stream (% by weight)[a]	Waste Loading (% by weight)[a]	Toxic Metal Concentration (ppm)				
		Pb	Cr	Cd	Hg	Ni
DOE Mixed Salts[b]	100	3,000	3,000	3,000	nt[c]	3,000
	60	0.07	0.10	0.37	nt	0.40
DOE Incinerator Ash[b]	100	5,000	5,000	5,000	nt	5,000
	60	0.99	0.25	1.42	nt	1.38
	60[d]	0.01	0.01	0.01	nt	0.18
WSRC Consolidated Incinerator Facility Blowdown[e]	100	2,250	500	125	250	500
	40[d]	<0.05	0.07	<0.05	<0.04	0.1
Idaho National Engineering Laboratory[f]	100	4.2	nt	nt	11.4	nt
	50	0.18	nt	nt	0.02	nt
Rockwell Molten Salt Oxidation Residuals[g]	100	nt	2,400	nt	nt	nt
	50	nt	1.6	nt	nt	nt
Current USEPA Maximum Allowable Concentrations		5	5	1	0.2	—[h]
Proposed USEPA Maximum Allowable Concentrations (USEPA 1995b)		0.37	0.86	0.19	0.2	—[h]

[a]Baseline TCLP leach data reported as 100% waste by weight
[b]Generic DOE Mixed Waste Surrogate based on data in Bostick et al. (1993)
[c]nt = not tested
[d] Including pretreatment additive
[e]Westinghouse Savannah River Consolidated Incinerator Facility (CIF) Blowdown Surrogate based on data in Kalb (1993)
[f] INEL Actual Mixed Waste sample (Lageraaen, Patel, and Kalb 1995)
[g]Rockwell Actual Mixed Waste sample (Lageraaen et al. 1995a)
[h]Not included in current US EPA definition of toxic characteristic metal

Table 3.17. Table 3.18 provides performance data from the TCLP for SPC-encapsulated waste. The Idaho National Engineering Laboratory fly ash (untreated) leachate concentrations were well above the TCLP-allowed concentrations. Leachability of the fly ash solidified with SPC was considerably lower, but was still above the allowable concentration levels. The addition of small amounts of sodium sulfide with SPC encapsulation reduced TCLP leachability of the SPC solidified fly ash below allowable concentrations. The sodium sulfide reacts with the metals salts to form metal sulfides that have low solubility within the solidified matrix (Kalb, Heiser, and Colombo 1991b).

Table 3.18
USEPA's TCLP Performance

Results from US EPA Toxicity Characterization Leaching Procedure
for INEL Incinerator Fly Ash Encapsulated in Sulfur Polymer Cement

Sample Tested (by weight)	Concentrations of Criteria Metals (mg/L)	
	Cd	Pb
Idaho National Engineering Laboratory Fly Ash	85.0	46.0
55% Ash 45% SPC	27.5	17.6
40% Ash 60% SPC	13.6	12.0
40% Ash 53% SPC 7% Na$_2$S	0.1	1.0
43% Ash 50% SPC 7% Na$_2$S	0.2	1.5
US EPA Allowable Limit	1.0	5.0

Source: Kalb, Heiser, and Colombo 1991b

3.2.14 Design Checklist

3.2.14.1 Ex-Situ Polymer Stabilization/Solidification

1. Characterize the waste stream to determine moisture content, particle size, radioactive and hazardous contaminant concentrations, and volatile organic concentration.

2. Determine waste pretreatment requirements (e.g., drying, size reduction).

3. Determine the waste constituents(e.g., mercury or volatile organics) that have the potential to be volatilized either during pretreatment or encapsulation processing. Develop an offgas system capable of capturing potentially volatile species.

4. Optimize polymer use based on waste characteristics, volume to be treated, performance goals, and costs. Consider substitution of recycled polymers for virgin materials.

5. Size system components to satisfy anticipated production requirements for treating as-generated waste as well as reducing the volume of stored inventories. Assess cost-effectiveness of using multiple parallel processing systems or a larger single unit.

6. Define the final waste form product performance criteria. Perform sufficient testing to confirm product durability and performance.

7. Define operating parameters to ensure cost-effective operation.

8. Develop process QA/QC (QAPP) and confirm through adequate pilot- and full-scale testing and implementation of advanced monitoring techniques.

3.2.14.2 In Situ Polymer Stabilization/Solidification

1. Characterize soil in terms of porosity, particle-size distribution, and contaminant concentrations.

2. Determine the waste constituents(e.g., mercury or volatile organics) that have the potential to be volatilized either during pretreatment or encapsulation processing. Develop an offgas system capable of capturing potentially volatile species.

3. Optimize the type of monomer based on the characteristics of the soil and waste, volume to be treated, performance goals, and economic considerations.

4. Select the catalyst and promoter based on compatibility with monomer and soil/waste characteristics. Conduct both bench- and pilot-scale testing to prove reliability of each.

5. Develop ratios of monomer, catalyst, and promoter based on consideration of processing (e.g., setting time) and cost.

6. Design system to avoid mixing components within delivery pumps, piping, nozzles, and other equipment which would be damaged by inadvertent setting of the polymer.

7. Optimize emplacement design to minimize waste of materials and ensure adequate coverage and resulting solidification of the contaminated soil.

8. Define operating parameters to ensure cost-effective operation.

9. Conduct test bores in situ to confirm material compatibility and QA/QC of the emplacement system.

3.3 Vitrification

3.3.1 Remediation Goals

3.3.1.1 Ex-Situ Melters

Vitrification technologies have the potential to be widely applied to hazardous and radioactive waste treatment. As an emerging technology, available performance data are largely based on pilot-scale demonstrations. A majority of demonstration and testing results have been funded by federal government agencies such as DOE, DoD, and US EPA. Demonstration tests have evaluated the technology for treating many types of wastes, such as soils contaminated with heavy metals and organics, asbestos-containing waste, industrial fly ashes and furnace dusts, sludges, and solids contaminated with radioactive or heavy metals.

The application of vitrification to high-level radioactive waste has been accepted worldwide as the preferred treatment option. Facilities in France, Russia, and Belgium have been operating since the late 1970's. Plants in the U.S. and Japan became operational in 1996 and 1997. These plants are expected to operate anywhere from 3 years to 30 years, depending on the specific mission at the respective sites. Microwave melting of mixed wastes has been demonstrated at the DOE facilities in Oak Ridge, Tennessee, and Rocky Flats, Colorado. However, large-scale production has not yet been implemented by the DOE.

Vitrification has been defined by the US EPA to be the BDAT for cadmium-bearing wastes.

3.3.1.2 In Situ Vitrification

In situ vitrification (ISV) is a versatile remediation technology that is capable of converting contaminated soil, buried waste, radioactive waste, and industrial and municipal waste to a solid waste form that satisfies waste disposal and site closure requirements. Offgas monitoring conducted during several major demonstrations has shown ISV to be effective in retaining, destroying, and/or removing contaminants to levels that satisfy permitted discharge limits.

At the Parsons Chemical Works, Inc., site in Grand Ledge, Michigan, for example, ISV was used to treat contaminated soil containing pesticides, mercury, and low levels of dioxins and furans (US EPA 1995c). The work was conducted in association with US EPA's SITE Program. The results showed that the vitrified waste met US EPA Region V cleanup criteria and that air emissions were below regulatory limits. A thermal oxidizer was used in the offgas train to destroy an odiferous (sulfur-related) nonhazardous offgas that had created public concern. Scrubber water from offgas treatment required secondary treatment. The work on the SITE Program test cell was completed in 10 days with only minor operational problems. During this time, approximately 540 tonne (600 ton) of contaminated soil were vitrified.

At a Toxic Substances Control Act (TSCA) demonstration project site, ISV was used to remediate 2,800 tonne (3,100 ton) of soil contaminated with varying PCB concentrations up to 17,000 mg/L (Thompson, McElroy, and Timmerman 1995). Some of the demonstration cells contained significant amounts of debris, including ruptured drums, asphalt, concrete, protective clothing, and other wastes. Dynamic compaction (see Section 3.3.5.2) eliminated major voids and breached drums containing water to prevent disruption of the melts. The results indicated that (1) 99.76% of the PCBs were destroyed by the melt; and (2) 99.98% of the entrained PCBs were removed by the offgas treatment system. Scrubber solutions containing PCBs were treated and disposed at a permitted TCD facility.

Geosafe Corporation recently applied the ISV technology to treat wastes at the Wasatch Chemical Superfund Site in Salt Lake City, Utah. Approximately 5,400 tonne (6,000 ton) of soil and debris contaminated with dioxin, pentachlorophenol, herbicides, pesticides, and other semivolatile organic compounds and VOCs were treated. Large amounts of debris, including scrap metal, plastic, wood, and clay pipe, were present within the treatment volume. The project was successfully completed in November, 1995. Samples of vitrified product, adjacent soil, and offgases indicated non-detectable levels of the contaminants of concern (see Table 2.2 in Section 2.3.2.2). A notable feature of this project involved the treatment of 650 gallons of oil contaminated with 11 mg/L dioxin. The oil was mixed with soil and then staged within the treatment zone. This was the first large-scale treatment of dioxin-contaminated waste within the U.S. using a non-incineration technology. The technology is not considered to be an incinerator, even though an afterburner was used for final destruction of organics. This is because greater than 99% of the organics are destroyed by pyrolysis rather than by oxidation.

At Oak Ridge National Laboratory, a pilot-scale test was conducted on a simulated waste trench containing small quantities of ^{137}Cs and ^{90}Sr (Carter, Koegler, and Bates 1988). The results of the test indicated that the durability of the glass waste form was similar to or better than glasses formulated for high-level radioactive wastes. However, a higher-than-expected quantity of ^{137}Cs was deposited on the walls of the offgas containment hood. High radioactive doses to personnel were projected at the observed deposition levels. Further testing is now underway at Oak Ridge National Laboratory to explore methods of safely using ISV when ^{137}Cs and ^{90}Sr are present. The testing includes use of a prefilter to remove ^{137}Cs before it enters the offgas treatment system. Preliminary results indicate that retention of ^{137}Cs in the melt was very high (99.9987%) and that 78.6% of the volatile ^{137}Cs was retained by the prefilter. The remainder of the volatile ^{137}Cs was found between the prefilter and the final filter.

3.3.2 Design Basis

To apply vitrification technologies, whether ex-situ or in situ, many of the same issues must be considered: waste composition and characteristics, resulting glass composition, degree of heterogeneity, offgas treatment requirements, and required production rate. Ex-situ treatment applications must also consider the volume of waste to be treated against the required treatment period. Additional considerations for ISV include certain site conditions, such as the depth of waste and host soil characteristics. Specific design issues for vitrification technology options are discussed in the following subsections.

3.3.2.1 Ex-Situ Melters

Electric Melters. Electric melter systems produce a glass product anywhere between 0.45 kg/hr (1 lb/hr) to 180 tonne/day (200 ton/day); with commercial glass melters representing the high end of the production rate spectrum. These melters are designed according to the required production rate, the composition and characteristics of the material to be processed, product requirements, and required operating life. Many of the waste streams under consideration for vitrification have never been processed before. For this reason, it is necessary to evaluate the factors discussed below.

Waste characterization should be completed first to allow initial glass development to be performed. This will also identify the relative concentrations of "troublesome" components that have limited solubility in glasses. These include sulfur, the halides, phosphate, and carbon. The initial glass work can be done with surrogates and will identify the glass system, e.g., sodium-aluminum-silicate, calcium-aluminum-silicate, borosilicate, or phosphate, that will provide the necessary product performance and best accommodate the troublesome components. Any glass-forming additives required to achieve a processable and high quality glass will be determined at this time. An estimate of the waste loading that can be achieved in the glass will also be made, often using laboratory- or bench-scale testing. Generally, higher waste loadings can be achieved with glasses with higher processing temperatures, such as wastes containing a high proportion of transition group oxides and refractory oxides like silicon, alumina, and zirconium.

Waste characterization is also important to estimate the degree of compositional consistency within the waste volume. It is preferable to establish a single glass composition and a fixed-glass former composition. When the composition is fairly consistent, the initial waste characterization and process sampling can be minimized. The size and complexity of the glass compositional field needed to ensure acceptable glass composition across the range of waste variability can also be minimized. Finally, the physical condition of the waste, such as water content, particle size, and viscosity (if a slurry), can be determined.

Once a glass composition is defined, melter construction materials should be evaluated. The nominal compositions and relevant property data of several refractories commonly used in electric melters are presented in Table 3.19. The glass industry predominantly uses an alumina-zirconia-silica (AZS) refractory in contact with the glass, because it will not color the glass as it slowly corrodes. Large commercial furnaces can be operated for up to two years before the wall refractories must be replaced. However, more frequent rebricking in areas of high glass flow, such as discharge throats, is required. Durability is improved through refractories that contain a high fraction of chromium oxide. The slow corrosion of these refractories results in small quantities of chrome oxide going into the melt. This has no effect on glass properties. Chrome oxide refractories are widely used in melters treating hazardous and radioactive wastes. Where insufficient experience exists, refractory vendors should be consulted to identify candidate refractories followed by laboratory corrosion testing to select preferred refractories.

Refractories which back up the glass contact refractory are selected for both chemical durability and insulative properties. The degree of heat loss that can be tolerated should be based on two factors. First, the glass temperature at the refractory wall should be high enough (above the liquidus temperature) to avoid crystal formation. The second factor is cost. Higher capital costs for initial melter fabrication reduces energy costs and outer shell water cooling costs during operation.

Refractories in the melter lid will be exposed to high temperatures during idling, extreme thermal cycling when feeding is started and stopped, and corrosive offgas constituents such as acid gases, salts, and feed and glass splatter. The selection of suitable lid refractories is based on the expected process conditions and required service life. Typically, high alumina castable refractories, firebrick, and insulating board are used.

Selecting the appropriate electrode material is also crucial. Graphite, tin oxide, and molybdenum metal are standard glass industry electrode materials that have been adopted for use in radioactive and hazardous waste

Table 3.19
Properties of Common Melter Refractory Materials

Refractory and Use	Al_2O_3	SiO_2	Cr_2O_3	ZrO_2	Fe_2O_3	MgO	CaO	Density (g/mL)	Conductance[a] (W/M • °C)
Glass Contact: Monofrax K–3[b]	60.4	1.8	27.3	—	4.2	6.1	—	3.9	2.9 (@1100°C)
Glass Contact: Monofrax E[b]	4.0	1.6	77.7	—	7.9	8.7	—	4.2	5.0 (@1100°C)
Glass Contact or Backup: AZS (e.g., Zirmul[c])	70.0	10.2	—	19.5	—	—	—	3.1	1.9 (@900°C)
Backup Insulating: Alfrax 57[b] Castable	94.8	0.4	—	—	0.1	0.1	4.7	1.2	0.7 (@550°C)
Backup Insulating: Alfrax 66[b] Castable	95.9	—	—	—	0.1	—	4.0	2.7	1.4 (@550°C)
Backup Insulating: Fiberfrax Duraboard[b]	56.0	39.0	—	—	—	—	—	0.3	0.07 (@300°C)

[a] 1 BTU • in./hr • ft² • °F = 0.144 W/m • °C
[b] A registered tradename of the Carborundum Co.
[c] A registered tradename of Chas. Taylor & Sons Co.

vitrification. Inconel-690 (INCO Alloys International, Huntington, West Virginia), a nickel alloy material, has been successfully used in high-level radioactive waste melters operated below 1,200°C (2,200°F). A chrome oxide ceramic has been evaluated recently for use in radioactive waste processing at temperatures up to 1,550°C (2,800°F) (Lamar, Cooper, and Freeman 1995). Inconel-690, as well as graphite and molybdenum have been used in electric melters that have been demonstrated for hazardous waste vitrification. Significant studies have been performed in the past two years to assess candidate electrode materials (Freeman, Sundaram, and Lamar 1995; Sundaram. Freeman, and Lamar 1995). The best electrode material is highly dependent on the glass composition and oxidation conditions that exist during processing.

The initial process testing can be performed with small engineering-scale equipment, typically about one-tenth scale. The test objective is to assess the basic processability of the feedstock. Chief measurements are processing rate, determination of any secondary phase formation, accumulation of unmelted phase deposition of hard-to-melt components, offgas composition, and extent of particulate, volatiles, and aerosol loss from the melter. The test duration should be long enough to ensure that the system has achieved steady-state operation and that the glass inventory in the melter has been turned over at least three times. This will result in the melter inventory, as well as the product glass, being approximately 97% derived from the feedstock and the remainder being the initial start-up glass composition. Testing determines whether the basic process as defined will provide acceptable treatment results. A pilot-scale system (one-fourth to one-third scale) is usually required to predict full-scale production rates and to obtain accurate offgas data. This is critical if the melter does not rely on mechanical or other means of agitating the glass batch.

Combustion Melters. Many of the design basis requirements for the electric melter are also requirements for combustion melters. These include waste characterization and analyses and laboratory glass development. Waste characterization and assessment of waste homogeneity are important because of the small glass holdup or inventory associated with combustion melters. With a small inventory, it is more important that the instantaneous feed composition closely approximate the target or nominal composition. Otherwise, glass quality could vary significantly. Troublesome components include sulfur, halides, and metals having low vaporization temperatures, such as arsenic, cadmium, and lead. The chief issue is to determine the fractional loss during processing of these components and the ability to recycle them back to the process for ultimate retention in the glass.

Refractories in combustion melters must withstand not only glass corrosion, but also high erosion, significant thermal cycling, and gas phase acid attack. Refractory selection depends on the combustion melter type. Some melters rely on refractories for thermal insulation in the combustion zone, while others use a cold-wall design and allow for higher energy losses. An added trade-off is the balance between refractory costs versus more frequent shutdowns for replacement of less durable refractories.

Combustion melters are noted for high production rates in relatively compact units. Therefore, testing can be performed at pilot-scale in reasonably small equipment. Process testing objectives are to establish production rate and obtain product quality data under varying conditions in order to define optimum operating conditions. Test parameters include feed particle size, feed composition variability, feed injection rate, fuel/air mixture, production rate, material losses versus combustion/melting chamber temperature, and chamber gas flow patterns. Because combustion melters can attain steady-state conditions in a matter of a couple of hours, many tests can be conducted in a short time.

Induction Melters. Many of the design basis requirements for the electric melter are also requirements for induction melters. These include waste characterization and analyses, and laboratory glass development. Waste characterization and assessment of waste homogeneity are important due to the smaller glass holdup or inventory associated with induction melters. Those components that cause problems for electric melters also present problems for induction melters.

Induction melters are constructed without refractories. Rather, a segmented wall constructed of water cooling channels forms the melter vessel. A schematic of an induction melter designed by Cogema is shown in Figure 3.19. The cold wall results in a 2 to 5 mm (80 to 200 mil) layer of glass being frozen at the wall. This protects the wall from glass corrosion. However, significant heat losses can occur, requiring significantly higher energy input than would an insulated melter. Above the melt line, the metal is exposed to corrosive offgases which can condense onto the cool surfaces. It is necessary, therefore, to consider this phenomenon during materials evaluations.

Figure 3.19
Cogema Induction Melter Schematic

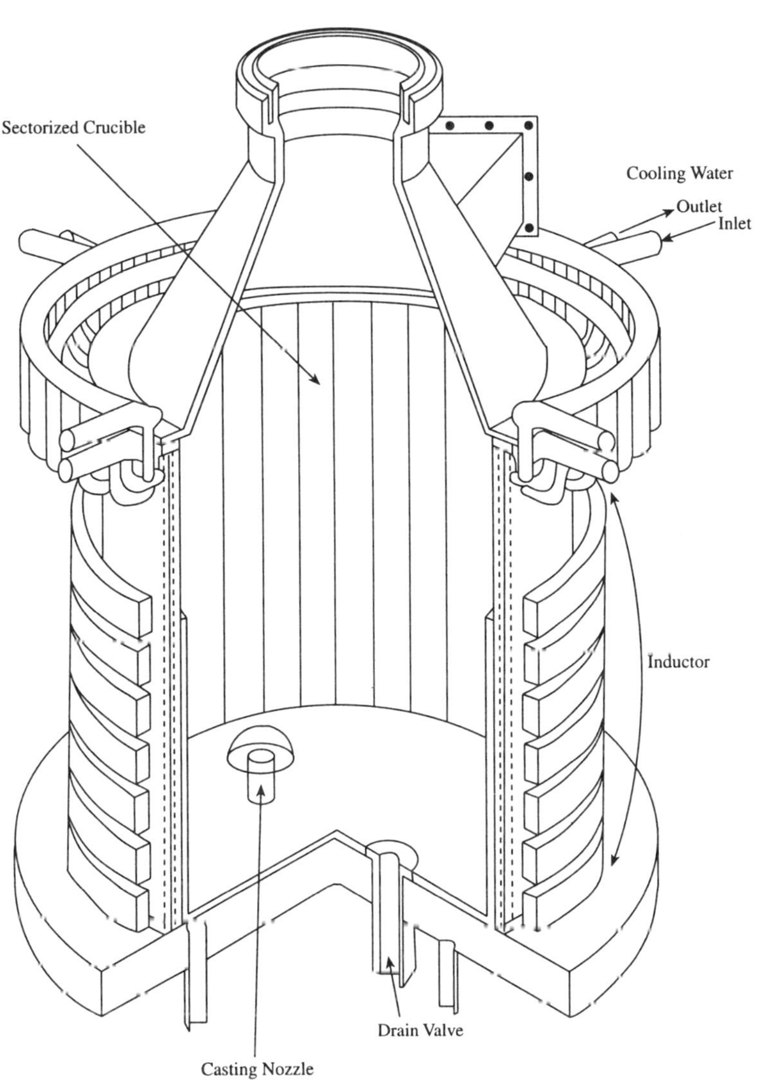

Reproduced courtesy of C.E.A. — Cold Crucible Melter (1997)

Presently, the largest high-temperature induction melter tested for treating hazardous or radioactive wastes has a melt surface diameter of 1 m (3.3 ft). The majority of testing and development has been performed in units ranging in diameter from 27 cm (10.6 in.) to 55 cm (21.7 in.). Initial process testing should be performed in such units. The primary test objectives are to determine process rate, effect of agitation on process rate, thermal efficiency, losses to the offgas due to entrainment or volatility, process characteristics, and rate as a function of feed characteristics. Regarding the latter objective, presently only dry feeding into induction melters is performed. Liquid feeding has not been established as a routine or recommended option. Therefore, feed materials should be relatively dry and reduced in particle size to optimize reaction and melting.

Microwave Melters. Microwave melting can be performed with less waste stream characterization compared to the other melters, since a heterogeneous waste stream can also be treated. However, because only the top 5 to 10 cm (2 to 5 in.) of material is in a molten state during processing, the heterogeneous characteristics of the feed are maintained in the final glass monolith. Requirements for a high quality product might impose additional characterization, blending, and sampling requirements.

Troublesome components that restrict the effectiveness of the microwave melter include strictly organic wastes and high metals content. Organics are substantially volatilized. High metals content can lead to localized arcing which reduces efficiency and, if next to the container wall, could melt a hole in the container. Significant concentrations of salts, halides, and high volatility metals, such as mercury, can be treated but could lead to poor quality products or high offgas losses.

Microwave melters vitrify the waste directly in the disposal drum. Therefore, no significant materials issues exist. The microwaves are conducted to the treatment drum through a wave guide. Isolation windows in the wave guide prevent material from migrating back to the generator.

Process testing can be performed with small- and full-scale units. Initial testing in small-scale (e.g., 4 to 20 L [1.1 to 5.3 gal] capacity) microwave units determines basic melting characteristics, waste and additive composition, and final product quality. Full-scale units employ 208 L (55 gal) drums as the melting/disposal container. Testing at full-scale determines achievable processing rates, optimum microwave energy input values, final product quality, and characterizes the offgas composition.

3.3.2.2 In Situ Vitrification

Each application of the technology requires sufficient knowledge of site conditions to support design of an effective system. The overall oxide composition of the test soil and waste determines key melting properties, such as fusion and melting temperatures, and melt viscosity. Soil to be treated must contain sufficient quantities of conductive alkalis (K, Li, and Na) to carry the current within the molten mass. Additionally, the soil should contain acceptable amounts of glass formers, for example, silica. Most soils worldwide have an acceptable composition for ISV treatment. Geosafe Corporation determines the oxides present in the soil prior to treatment. A computer-based model then determines the suitability of the site for vitrification. The model can also identify solids and wastes that require modification before treatment (US EPA 1994a).

The type of contamination present on-site affects the design of the offgas treatment system more than it affects the design of the rest of the ISV system. For this reason, the offgas treatment system is modular in configuration, allowing treatment of the offgases to be tailored to site-specific conditions.

The limited ability of the current offgas equipment to remove heat dictates that the organic content of the treatment media be less than 7 to 10% (by weight) thereby allowing for adequate offgas quenching temperatures. Very high metals content (estimated >50% by weight) and inorganic debris (estimated >75% by weight) may be treated by ISV as long as the arrangement of such materials within the soil matrix allows satisfactory performance.

Previous experience has indicated that safe, effective treatment cannot be assured if large voids and/or pockets of liquids in sealed containers exist beneath the soil surface. The gases released can cause excessive bubbling of molten material, resulting in a potential safety hazard. For this reason, sufficient site characterization is recommended prior to treatment if buried drums exist or are suspected to exist beneath the soil surface. Uncontained combustible materials generally do not cause processing difficulties since they decompose relatively slowly as the melt front approaches. Full-scale demonstrations have been successfully conducted on sites containing significant quantities of combustibles, such as wooden timbers, automobile tires, personal protective equipment, and plastic sheeting.

The presence of large amounts of water in the medium requiring treatment can hinder the rate of ISV, since electrical energy is initially used to vaporize this water instead of melting the contaminated soil. The resulting water vapors must also be handled by the offgas treatment system. Treatment times are thus prolonged and costs increased when excess water is present. If ISV will be performed at depths below the water table, and the hydraulic conductivity of the surrounding media exceeds $4 \cdot 10^{-4}$ cm/sec, it might be economically advantageous to use some means of dewatering or limiting the water recharge rate.

The maximum acceptable treatment depth with the current equipment is 6.1 m (20 ft) Below Ground Surface (BGS); however, full-scale tests at Geosafe's testing facilities in Richland, Washington, have demonstrated that existing large-scale equipment can successfully melt to a depth of approximately 6.7 m (22 ft) BGS. Melts at the Parsons site typically reached depths of 4.6 to 5.8 m (15 to 19 ft) BGS.

Site conditions should also be evaluated for the soil's ability to support a forklift and a 125-ton crane used for changing containment hood positions. Gravel, timbers, or other material can be used when the load-bearing characteristics of the soil are not adequate. Sufficient space for maneuvering the crane and positioning equipment is also required.

3.3.3 Design and Equipment Selection

Ex-situ and in situ melters can generally be considered as single unit systems and, therefore, are fairly simple in design compared to more mechanically complex, S/S processes. The electric melters are designed around standard 480V/60Hz/3ø power sources. Depending on the melter and its design, the process engineer specifies transformers and power controllers to obtain the voltage, current, power, and phase required for the system. The following subsections describe additional information needed to design the technology.

3.3.3.1 Ex-Situ Melters

Electric Melters. Sizing of electric melters which do not use agitation is based on glass surface area. This value is strongly affected by the waste stream being treated. High-level radioactive waste slurry containing about 50% (by weight) solids can be processed at about 960 kg/m²-day (200 lb/ft²-day). Dry solids, such as soils, can be processed at rates 50 to 100% higher. Production rate also depends on the individual waste constituents. For instance, high concentrations of nitrates, carbonates, or organic materials generate large quantities of decomposition gases which can interfere with the melting process by inhibiting heat transfer. Therefore, it is important to obtain sufficient pilot plant data to support melter sizing.

Commercially-available melter systems have incorporated active mixing using mechanical or pneumatic (i.e., gas bubblers) means. These systems, such as the Stir-Melter® system shown in Figure 3.20, negate some of the effect composition has on melter sizing. Stir-Melter, Inc. has stated that they can process equivalent rates in melters just 10 to 15% of the size of unagitated melters. However, verification testing is still necessary to properly size and specify a unit. Numerical modeling is typically used to specify refractory thicknesses and placement to meet thermal and mechanical stress requirements. These models can also analyze electric potential fields and steady-state convection flow fields as functions of electrode placement, power levels, temperature, and glass properties (e.g., viscosity and electrical conductivity)(Eyler et al. 1991).

Power, voltage, and amperage requirements depend on melter size, glass properties, and melter geometry. As a rule of thumb, it is assumed that roughly one kilowatt-hour of power will be required for every kilogram of glass produced. This assumes little or no usable heat content exists in the waste stream. Glass resistivity combined with melter dimensions will establish the electrical resistance by the following relationship:

$$R = \varepsilon(l\,/\,A) \qquad\qquad (3.1)$$

where: R =   electrical resistance across electrodes in ohms;
             ε =   resistivity of the glass in ohms/cm;
             l =   distance between electrodes in cm; and
            A –   area perpendicular to electrodes in cm^2

The minimum electrode surface area is defined by the maximum current flux recommended to prevent "destructive heating" of the glass at the electrode-glass interface (Yellow Book). It has been recommended that current fluxes be maintained less than 2 amp/cm^2 (13 amp/in.^{2}) to prevent significant electrode consumption rates. Transformers, power supplies, and controllers designed to operate at low voltage (e.g., 100 to 200 V) and high amperage (e.g., 1,000 to 3,000 amp) are more economical. Therefore, melter geometry and glass resistivity should be optimized according to these parameters.

Figure 3.20
Stir-Melter, Inc. Agitated Melter Schematic

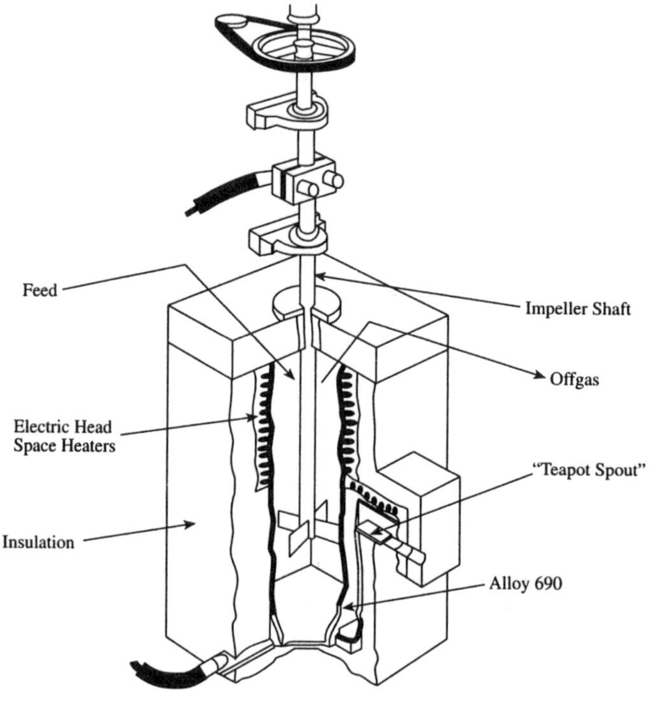

Reproduced courtesy of Stir-Melter, Inc.

Combustion Melters. No significant information has been identified in the literature to describe in detail the design process for sizing combustion melters. It is assumed that proprietary engineering and empirical data resides with the developers of the technology to support scale-up calculations. In general, melter size is a function of the residence time required to heat, react, and melt the waste stream. Two combustion melter systems, the Vortec cyclone melting system shown in Figure 3.21, and the Babcock Wilcox cyclone furnace (Czuczwa et al. 1993) have been subjected to a significant number of tests to demonstrate their processes in waste treatment.

Figure 3.21

Schematic Layout of Vortec Corporation's Cyclone Melting System

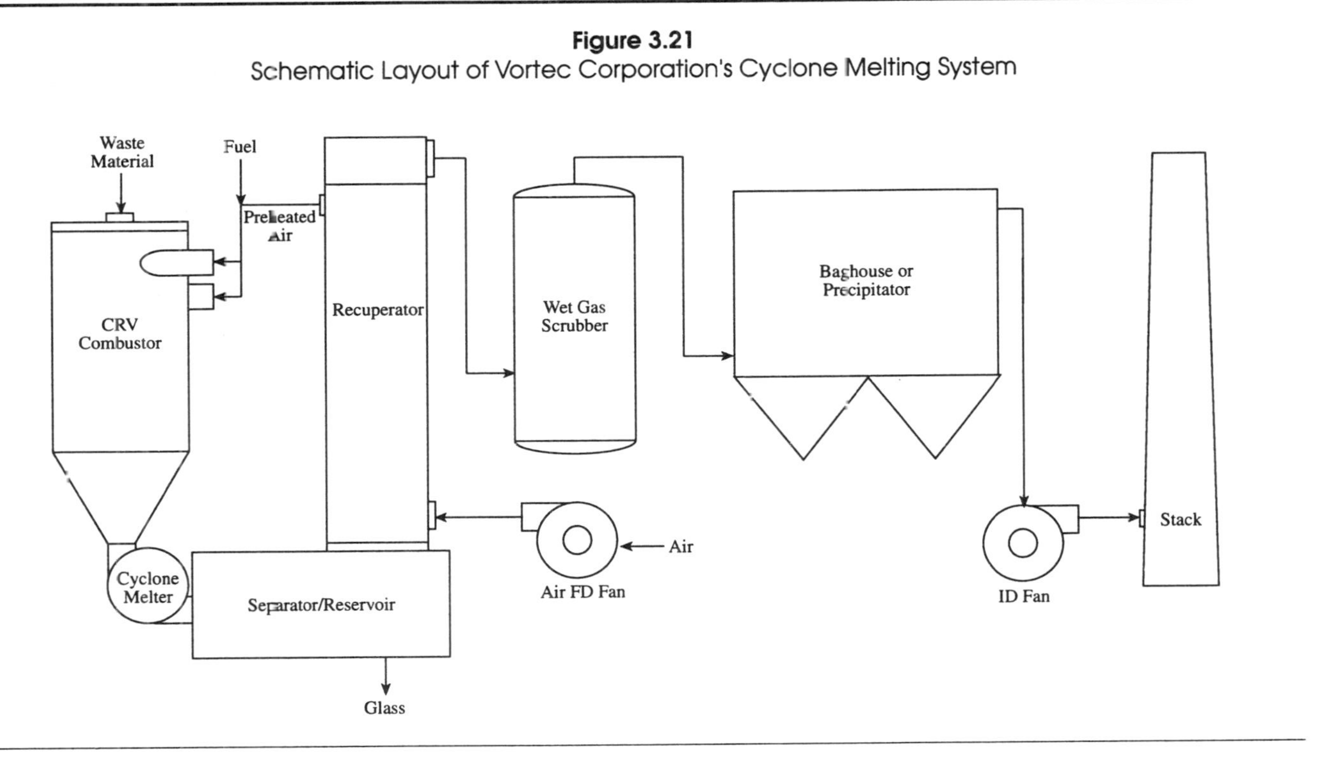

Both systems were originally developed as compact, high-efficiency thermal combustion units for supplying heat to boilers or processes. As such, significant expertise in radiant and convective heat transfer analysis and modeling should exist. This understanding, combined with pilot-plant tests, would be used to design melters for waste treatment.

The selection of fuel source — natural gas, coal, or fuel oil — determines design of the fuel and combustion air supplies. No strong emphasis has been placed on one fuel source over another. For this reason, economics may be the major factor. However, the presence of inert substances in coal, particularly heavy metals and sulfur, must be considered to ensure that offgas treatment and glass product quality are not affected to an unacceptable degree.

Induction and Microwave Melters. Because of the compact size of induction and microwave melters, design and equipment selection issues are few. The key parameters are the characteristics and composition of the waste material. Equipment suppliers will most likely perform testing in smaller units and use these results to specify the commercial unit. Because these units process less than 100 kg/hr (220 lb/hr) they are most applicable to limited waste volumes or in cases for which multiple process lines could be used. Induction melters are only now being built and tested with melters up to 1 m (3.3 ft) inside diameter. As melters increase in diameter, it is necessary to employ lower frequency ranges so that penetration into the bulk of the glass occurs. Dryer or calcination furnaces should be considered as a process step prior to the induction melter when treating liquid waste streams. For solid waste streams, calcination should be considered as a process-boosting step. Microwave melters can accept wet waste streams and even slurries. However, overall microwave production rates can be enhanced by dewatering and drying.

3.3.3.2 In Situ Vitrification

A predictive model of the ISV process has been developed at Battelle Pacific Northwest National Laboratory to assist engineers and researchers with the application of ISV at different sites (US EPA 1992a). The model, configured on a personal computer, predicts vitrification time, melt depth and width, and electrical consumption. Predictions are based on data inputs of electric operating parameters, soil parameters, and molten-glass characteristics. The model's predictions are useful for planning operations, cost estimating, and defining melt locations. The depth and width predictions, for example, can be used to locate the melts to help ensure that the entire contaminated region is treated and that adjacent structures are not damaged by ISV treatment. The model has been used to predict melt time, melt shape, and energy consumption; the model has accurately predicted monolith shapes of a large-scale ISV melt.

Luey et al. (1992) also reported that the important modeling parameters appear to be scale, power level, power-to-surface area ratio (governed by power level and/or electrode spacing), and, to a lesser degree, electrode diameter-to-spacing ratio. As the scale increases, the percentage of bottom heat transfer and relative downward growth decreases. Increasing the power supplied to the electrodes tends to produce a hotter melt region, higher flow speeds, promotes mixing, and consequently enhances the percentage of total heat loss passing through the bottom and the sides of the vitreous zone.

Power supplied to the system (typically about 4 MW for a large-scale unit) is limited by current and voltage limits imposed on the equipment. Approximately 700 to 1,000 kWh are consumed for each ton of soil vitrified. The system requires 12.5 or 13.8 kV, three-phase electricity. Electrode diameter (typically 30 cm [11.8 in.] for a large-scale unit) must be sized for the power level to prevent excessive electrode temperature and oxidation. Electrode spacing depends on the electrical conductivity of the melt and other factors, and typically varies between 3.4 and 4.6 m (11 and 15 ft), depending on site-specific conditions. A 3.5 m (11.5 ft) spacing was recommended for a proposed large-scale test at the Idaho National Engineering Laboratory. The gas containment hood is designed to contain potential bubbling of molten glass and to direct offgases to the offgas treatment system. Seals are created between the hood and the surface soils, and the hood and electrodes, to restrict air in-leakage to about 99% of the total gas collected at a negative pressure of about 1.3 cm (0.5 in.) water.

Offgas is drawn from the containment hood at about 200°C (392°F) and is usually cooled in a quencher. The quenched offgas usually is then scrubbed in one or more scrubbers to remove particulates and some gases. The design of the offgas treatment system depends largely on the contaminants expected in the offgas and their concentrations. Geosafe's large-scale unit can process 1,800 scfm (51 m³/min) resulting in CO emissions below 0.5 g/hr (0.001 lb/hr), NO_x emissions below 500 g/hr (1 lb/hr), and particulates below 9 g/hr (0.02 lb/hr). The thermal oxidizer consumes about 880 kW of natural gas when in use. The on-line efficiency of the overall system is 83 to 90%, according to Geosafe.

The time to mobilize or demobilize the equipment at a site is two to three weeks. A typical melting time is 10 days when processing 3.5 to 5.5 tonne/hr (4 to 6 ton/hr). The time required to move a containment hood and connect electrodes at a new setting is about two days. Using two hoods, Geosafe has been able to perform the power-down/power-up transition between melts in less than 12 hours.

3.3.4 Process Modifications

3.3.4.1 Ex-Situ Melters

Most potential modifications are driven by the desire to increase efficiency and production rate, or accommodate materials not generally thought to be processable. Electric melter technology is based on the processing of inorganic materials, i.e., glass making materials, such as silica, boric oxide, lead oxide, sodium carbonate, calcium silicate, etc. Process modifications would necessarily extend this basic application to the complex physical and chemical characteristics of hazardous and radioactive wastes. The following design or operational modifications can be considered for ex-situ melters:

- increase processing temperature to achieve higher waste loadings and higher production rates. Higher cost and consumption rate of electrode materials can occur; as well as an increase in the rate of volatility of halides, alkalies, high vapor pressure metals, etc.;

- add supplemental heating in the plenum space of a melter to boost production rate and maintain space temperature above 1,000°C (1,800°F) to assure combustion of organics. Gas, electric, or plasma heating have been demonstrated;

- increase plenum space volume to accommodate wastes with a high organics content and promote pyrolysis, combustion, and decomposition of organics;

- incorporate mechanical, gas sparging, or other means of mixing the glass to increase heat transfer to the waste, thereby increasing the production rate and reducing the relative size of the melter;

- introduce oxygen or air jets into the plenum space to aid in oxidizing organics (assuming the air is mixed with the gases containing the organics at sufficient temperatures; a separate afterburner might be the best choice);

- introduce oxygen or air jets into the melt to aid in oxidizing metals; and

- for wastes containing high concentrations of metals, a melter should be designed to accumulate and drain a molten metal stream separate from the glass stream. For electric melters, this affects electrode design and placement, and refractory linings should be replaced with a cold wall design. For induction melters, multiple zone heating would need to be designed to account for the difference between glass and metal induction properties.

3.3.4.2 In Situ Vitrification

Applications involving buried waste might result in variable offgas evolution rates and excessively high offgas heat loadings if a high fraction of combustible materials are present. Such applications might require preconditioning of the wastes, prior to ISV processing, to avoid or minimize these effects. Site conditions that may warrant preconditioning include:

- lack of necessary chemicals in the soil to ensure uniform melting and attainment of a suitable vitrified product;

- excessive void spaces;

- excessive liquid quantities or liquids in sealed containers;

- excessive heat generation related to the oxidation of the products of pyrolysis of waste components; and

- significant uncertainty regarding the materials and conditions present. Such conditions may render the direct application of ISV more challenging for a variety of technical, economic, and safety reasons.

In most cases, engineering means are available to effectively deal with these concerns. The primary engineering means include:

- adding soil with appropriate chemical additives on the surface above the waste materials;

- modifying the chemistry of the contaminated media by injecting soil or chemicals into the volume to be treated;

- densifying the waste site using dynamic compaction (see Section 3.3.5.2);

- rupturing sealed containers using dynamic disruption (see Section 3.3.5.2);

- removing liquids using thermally-assisted vacuum extraction;

- filling voids by injecting solids; and

- excavating and removing unacceptable materials such as explosives, and then re-staging the wastes in a manner acceptable for ISV processing.

The processing rate can be increased using multiple ISV units or a second offgas containment hood. The use of a second hood enables the hood and electrodes to be installed at a new setting while ISV is being completed at a current setting. Battelle has also studied the feasibility of significantly larger units as a means of increasing processing rates and concluded that units two to three times the size of the present unit can be built and operated (US EPA 1995c).

3.3.5 Pretreatment Processes

3.3.5.1 Ex-Situ Melters

Industrially-supplied equipment should satisfy the handling, preparation, and feeding of materials to the melter for most applications. For liquid or slurry systems, properties requiring definition include composition (to assess corrosion potential and material hardness requirements), percent dissolved and undissolved solids, particle-size distribution, viscosity of the mixture, and requirements for homogeneity. Selection of equipment for size reduction and conveyance or pumping should be based on capacity requirements and physical and chemical properties of the feedstock.

When the waste material is received at the processing facility, a "macro" batch should be characterized to determine the amount of glass or chemical additives. To assure the sample is representative, mixing equipment should be used to dry or wet blend the material. Size reduction or classification equipment should also be used at this point if the waste stream contains rocks or other foreign matter. The material should be crushed or sorted and the foreign material and large rocks removed and discarded or treated separately. Size reduction is also necessary to increase the surface area of the material so that it will react and melt more quickly in the melter.

Thermal treatment systems, such as melters, are not efficient evaporators. If a waste stream has a significant free water content, the cost benefits of installing a dewatering or evaporation unit prior to the melter unit should be analyzed. Equipment options to consider include mechanical dewatering, thermal drying, microwave drying, and calcination spray or rotary furnaces. Costs for the added equipment would be weighed against the benefits of a smaller melter unit and reduced energy costs, since thermal efficiency should be higher in an evaporator. The complexity or selection of offgas treatment equipment in the absence of a large water load should also be considered. As was stated in Section 3.3.3.1, induction melters require removal of free water. Radioactive production units have been designed to couple the dryer or calcination units directly to the melter. A similar design configuration might be optimal for hazardous waste treatment applications.

Prior to feeding a melter, any necessary glass or chemical additives (as determined by the waste characterization) should be blended with the waste stream. These additives would be stored in large bins from which the required amount can be automatically transferred to a blending bin and then blended with the waste. Pre-blending is an important step when using a melter that does not have a means to mix the material once it has been fed. If pre-blending does not occur, the waste and the additives can separate and impede the melting rate.

Finally, solid waste and bulk chemical streams can generate appreciable amounts of suspendable solids. Liquid wastes can contain low vapor pressure organics that must be excluded from entering the workspace environment. The design engineer has to consider these aspects and identify appropriate ventilation, filtration, and treatment systems.

3.3.5.2 In Situ Vitrification

In some cases, it is necessary to add chemicals or soil of specific oxide composition to the treatment zone to obtain the desired properties of the melt and vitrified product. Materials can be added on top of the materials to be treated. In such cases, the melt is initiated in the added soil layer, and the desired composition is attained when the target contaminated material melts and mixes with the molten surface material.

It is possible to overcome deficiencies in chemical composition by injecting the required materials directly into the volume to be treated (Luey et al. 1992). This option might be preferred if large voids exist. The materials to be added should, in general, have a significantly lower melting point than the waste medium to be treated.

Large void volumes are of concern in ISV applications because they hold the potential to drain off a significant quantity of the melt, resulting in the loss of electrical continuity between electrodes. Moreover, release of the gas present in large voids can cause undesirable melt disturbances when the gas rises through the melt. In such cases, compaction can reduce the size or eliminate large voids.

Dynamic compaction is a proven method of densifying soils. This method involves dropping very large weights from heights of about 15.3 m (50 ft) onto the surface of the treatment zone, thereby compacting it. Another method, known as dynamic disruption, involves vibrating a vertically oriented I-beam or similar structural member from the surface down through the treatment zone on a spaced grid to ensure that the affected area has been conditioned. Typically, dynamic disruption creates a 0.9 to 1.5 m (3 to 5 ft) radius of influence. Materials present within the radius of influence are shaken, compacted, and disrupted. Sealed containers within the treatment zone are damaged and lose their sealing integrity.

High concentrations of organics (>10% by weight) can generate heat loads that exceed the ability of the offgas treatment system to adequately quench the offgas temperature. An excessive heat load could cause the equipment to malfunction or otherwise not meet performance specifications. Excessive heat loads can also cause overheating and damage the offgas collection hood and treatment equipment, although refinements in the hood design and operating procedures in the past several years have reduced the potential for damage.

The primary method of controlling the rate that heat is generated by the oxidation of pyrolysis products involves controlling the power applied to the electrodes, which in turn, affects the rate of melting and the corresponding rate of melt movement through the soil and combustible waste. For buried waste applications in which significant void volumes exist, the site should be compacted, or the voids should be filled by injection of solids to ensure proper control of melting rates.

Another method of controlling the rate of heat generation is to mix soil that is highly contaminated with organics with soil with low organic concentrations. Clean soil can be added if needed to attain acceptable organic concentrations.

If a site cannot be made acceptable for ISV processing using the pre-conditioning options presented above, then re-staging the buried wastes for treatment should be evaluated. Re-staging wastes for treatment consists of the following steps:

- excavation;

- removal of unacceptable materials or conditions, if any;

- crushing, grinding, or shredding oversize materials, as necessary, to improve ISV processing; and

- placement of the remaining materials within the soil in an acceptable manner for ISV processing.

Such re-staging is advisable when the site characterization information is inadequate to support decisions regarding pre-conditioning options. Re-staging can also be required if potentially explosive materials (e.g., unexploded ordnance, pressurized gases, combustible materials present with oxidizers, etc.) exist within the site.

3.3.6 Posttreatment Processes

3.3.6.1 Ex-Situ Melters

The glass or glass and crystalline product of vitrification can be either cast into monoliths using containers such as 208 L (55 gal) carbon steel drums or converted to a cullet and handled as a bulk material. One of the primary benefits of thermal stabilization/solidification treatment is that the product has very good durability. If necessary, the glass composition can be designed to ensure that the product will pass the TCLP test. Its subsequent disposal or use depends on the original waste stream and US EPA regulations governing the waste stream. Handling the product as a cullet eliminates the cost for the drums. To produce a cullet, the molten stream can be either poured into a water trough or it can be poured through water-cooled rollers or similar devices designed to form small glass pieces.

The offgas stream treatment requirements depend greatly on the waste stream. Typically, five components of the offgas stream must be considered. They are moisture, particulate, soluble gases, noncondensable gases, and non-combusted organics.

- Moisture can be condensed from the offgas stream or passed though the treatment train and allowed to go out the stack. If condensing the water is required, quench spray systems such as venturi scrubbers are typically used. If the water does not need to be condensed, the engineer must pay attention to ensure that the offgas stream temperature remains above the dew point throughout the treatment train to prevent condensation. If condensation and recovery are necessary, particulate matter and condensable and soluble gases such as HCl, and some fraction of noncondensable gases, such as SO_x or NO_2, are also co-recovered along with the water.

- Particulate matter entering the offgas stream can constitute 0.1% or more of the material being fed to the melter. It will include gross particulate with particle diameters greater than 10 μm. It will also include submicron material that has volatilized from the glass surface and recondensed in the offgas stream. Both wet and dry particulate scrubbers are commercially-available to remove particulates. Common dry removal processes include fabric filters, e.g., baghouses, electrostatic precipitators, and cyclone separators. Wet scrubbing devices include venturi and free-jet scrubbers and wet electrostatic precipitators. If necessary, polishing scrubbers can be placed downstream of such devices to capture any submicron particles that escaped the primary scrubber. These devices include high-efficiency fiberglass filters and High-Efficiency Particulate Air (HEPA) scrubbers.

- Water soluble gases are predominantly halide acid gases. There are many commercially-available treatment methods that can be applied to halide acids.

- NO_x and SO_x are typically the primary noncondensable gases produced from waste processing when nitrate and sulfate containing wastes are processed. Both of these are partially water soluble. Wet and dry scrubbing technology exist for SO_x recovery. NO_x scrubbing can be performed wet or with selective catalytic conversion to nitrogen and oxygen.

- Organics can be effectively oxidized if held at a sufficiently high temperature in an oxidizing environment. Typically, a secondary combustion chamber is used. However, due to the cost of activated carbon, it is not normally employed unless the gas volume and temperature are both low.

It should be remembered that besides the individual constituents in the offgas stream, the composite composition and the absolute concentrations of the constituents in the stream are of equal importance. This affects both the sequencing of treatment devices and the selection of the devices themselves.

3.3.6.2 In Situ Vitrification

The major process residual of ISV operations is the vitrified waste monolith. The glass and crystalline vitrified ISV product has been shown to be highly durable and resistant to leaching (Luey et al. 1992). It is possible that areas that have not been adequately melted between melt settings might be revealed during quality assurance evaluations of the monoliths. Core drilling of the monolith can be used to verify waste treatment depths and product consistency. Unmelted areas can be retreated using ISV and, where appropriate, alkalis and glass formers can be injected to facilitate melting.

A number of secondary process waste streams are generated by the ISV technology. These include air emissions, scrubber slurry, decontamination liquid, spent carbon filters, spent scrub solution bag filters, spent HEPA filters, used hood panels, and discarded personal protective equipment (US EPA 1994a). The amount of scrubber slurry and filter waste generated depends on the nature of the contaminated media before treatment. High particulate loadings in the offgas and high soil moisture contents can increase the quantities of these wastes. The number of used hood panels to be disposed depends on the corrosiveness of the offgases generated during treatment (as well as the corrosion-resistance of the hood panels) and the temperature and duration of treatment (US EPA 1994a).

Some process residuals (e.g., used scrub solution bag filters, spent HEPA filters, and discarded personal protective equipment) can be disposed by vitrifying in subsequent ISV settings to reduce the amount of these wastes that require disposal off-site. Geosafe vitrified all process residuals in subsequent settings at the Wasatch Site. Scrubber slurry generated during treatment may require special handling, depending upon the types and level of contaminants removed in the scrubber (US EPA 1994a).

Typical treatment processes for scrubber slurry include activated carbon adsorption, neutralization, precipitation, coagulation, flocculation, and ion exchange. Granular activated carbon filters can be regenerated and reused. Metal HEPA filters, if used, can also be cleaned and reused.

3.3.7 Process Instrumentation and Controls

3.3.7.1 Ex-Situ Melters

Electric Melters. Electric melters are controlled by adjusting electrode power based on temperature measurements of the melter tank. The glass can be measured directly using thermocouples placed in the glass, electrodes, or refractory. Based on these readings, the electrode power can be adjusted accordingly. The temperature is also measured in the glass discharge area, the plenum space, cooling water or air circuits, and in strategic areas within the refractory, if desired. Electrical measurements include electrode amperage, resistance, voltage, and power; similar measurements are taken for any resistance heaters used in the glass discharge area or plenum space.

Pressure values are measured in the plenum space and pressure and flow data are measured throughout the offgas treatment line. It is typical that a slight negative pressure (e.g., 5 to 13 cm [2 to 5 in.] water), be maintained in the melter plenum space to prevent offgases from leaking out of flanges into the general work area. Pressure and flow readings in the offgas treatment line between equipment pieces are used to verify steady-state operation and identify equipment or line sections that are developing a blockage.

Process flow rates of the feed to the melter and glass product are used to measure production rate stability. Visual observation of the feed pile on the glass surface, either through a view port or camera is still a key means to evaluate processing efficiency and whether the melter is being overfed or underfed. As experience with a particular waste stream is gained, production efficiency can be inferred by plenum temperature and pressure readings, and glass temperature, power, current, and resistance readings. Flow rates of the cooling services are also required to ensure they are maintained at set values. All of these data can be collected using computer interface equipment for subsequent inspection, data archival, or for feedback control.

Combustion Melters. This process requires measurements similar to electric melters. Noticeable exceptions are the measurement of fuel flow rate to the process, combustion air temperature and flow, oxygen concentration in the exit gas stream, and temperature in the combustion chamber.

Induction Melters. Induction melters are very similar to electric melters. Unique instrumentation includes induction power generator readings instead of electrode readings, and coil water flow and temperature and cold-wall cooling circuit flow rates and temperature.

Microwave Melters. Microwave melters are the least complex of all the melters. Monitoring and control mechanisms are limited to the microwave generator power level, microwave leakage detection, generator cooling water flow rate and temperature, and container and product temperatures, which are monitored using thermocouples or pyrometers.

3.3.7.2 In Situ Vitrification

The ISV process instrumentation and control system includes approximately 50 separate instruments used for tracking temperature, pressure, amperage, voltage, flows, gas concentrations, and other parameters. The system automatically logs all parameters and provides visual displays of data trends. Operators evaluate trend data against operating limits established in the permit and work plan as the primary basis for ISV process control decisions.

The ISV system is manually controlled by operators with the exception of two control valves that control the flow of air into the hood and the offgas system to maintain adequate negative pressure in the offgas hood. Automatic systems are used for certain important functions. For example, in the event of power outage, the system automatically starts a diesel generator that powers a backup offgas treatment system. The cost of ISV is minimized by operating as close to design and permit limits as possible. Manual control of the system allows for fine-tuning to optimize process performance.

3.3.8 Safety Issues

3.3.8.1 Ex-Situ Melters

All ex-situ melter types have safety requirements because of the use of high voltage or current and high temperatures. Electrical components should be specified, designed, and installed according to National Safety Code specifications. Safety requirements should include:

- using enclosures with positive-closure latches for exposed electrical sources, such as bus bar connections, fuses, and power supply wire connections;

- implementing a lock-and-tag procedure to isolate energy sources and protect workers;

- using personal safety equipment, such as rubber mats and rubber-lined gloves when inspecting or troubleshooting energized systems;

- using water cooling, insulation, or screening on hot surfaces to prevent contact;

- using Personal protective clothing such as high-temperature clothing, face shields, and insulated gloves for use when sampling the molten glass stream, probing the melter tank, or handling the hot product; and

- observing OSHA and National Industrial & Occupational Safety & Health (NIOSH) requirements governing industrial plant worker safety.

Combustion melters will use either natural gas, fuel oil, or coal. There are special handling, monitoring, and storage requirements for safety using these fuels. Microwave melters must be monitored for microwave leakage during operation with a hand-held detection device.

If the normal ventilation system fails due to mechanical or operational reasons, the hazardous gases, superheated gas, and steam generated during processing must be discharged through an emergency vent outside the building. Otherwise, the melter would pressurize and a large amount of glass could be discharged or the gases would leak into the workers' area from flanges or fittings.

If a waste stream being treated is a designated CERCLA or RCRA waste, the operating personnel must receive 40 hours of Hazardous Waste Health and Safety Training and 72 hours of supervised on-the-job training to comply with US EPA and OSHA requirements, along with 8 hours of refresher training.

3.3.8.2 In Situ Vitrification

The design of the ISV system is based on OSHA, the National Electric Code, and other applicable safety standards. An external water spray system installed outside the offgas hood ensures that the hood does not overheat if excessive bubbling of glass occurs. This condition can expose the hood to high radiant heat or high gas temperature if high fractions of combustible materials are being processed. Fencing and other physical barriers are installed to prevent trespassing. Administrative controls and training are used to ensure operator safety. For example, power to the electrodes must be disengaged and locked out when it is necessary to perform work on the offgas hood.

All components are connected to a common electrical grounding grid consisting of six-foot long copper rods driven into the ground and electrically connected. Personnel who require access to the site must complete 40 hours of Hazardous Waste Health and Safety Training and subsequent refresher training.

3.3.9 Specification Development

3.3.9.1 Ex-Situ Melters

Melter equipment components are used routinely by industry, and are currently designed to meet rugged industrial standards. Manufacturers and suppliers of refractories, electrical transformers, power controllers, and solid and bulk handling equipment should be contacted early in the design process to discuss the application to identify any special requirements. For instance, refractories can be produced with a lower volume fraction of voids or cut and finished with different surface smoothness criteria than are standard. However, additional delivery time and costs are likely to be incurred.

Depending on the waste stream and location, specifications might need to be included to protect electronic or electrical equipment from significant dust loadings in the air. Also, given that the equipment might process a variety of different waste streams, a robust design that can be easily disassembled and reassembled following inspection and repair should be an objective of the process design engineer.

3.3.9.2 In Situ Vitrification

All equipment must meet applicable safety and performance standards to mitigate the electrical, mechanical, and thermal hazards associated with the ISV system and the chemical, physical, and radiological hazards associated with the hazardous waste site. The ISV design specifications must be based on limits defined in the operating permit and conditions of the site. Key conditions include organic content, metal content, void sizes, moisture content, electrical conductivity, and depth of contaminated media. The types and concentrations of contaminants present impact the type of offgas system that must be used. For example, a site containing only heavy metal contaminants would not require thermal oxidation of the offgas, although scrubbing, gas filtration, and/or other treatment units may be required to ensure compliance with permit conditions.

Geosafe Corporation and ISV Japan, Ltd. are the only licensed vendors of ISV, and the two entities are business partners. Geosafe's scope is worldwide, whereas ISV Japan operates only within the boundaries of Japan.

3.3.10 Cost Data

3.3.10.1 Ex-Situ Melters

There are no known operating commercial treatment plants that have published reliable cost data. The following information is based on studies, US EPA SITE demonstrations, or vendor literature. A possible application of electric melter technology to treat contaminated soils, sludges, and debris was evaluated by Koegler et al. (1989). Capital cost for a 180 tonne/day (200 ton/day) melter, power supplies, feed handling, offgas treatment equipment, and glass handling equipment was estimated to be \$5.5 million. Facility design, construction, and equipment installation were approximately twice the capital equipment cost. For this application, electric power cost, was 6¢/kWh and comprised 57% of the total estimated operating cost of $60 million. Labor cost was only 7% of the operating costs.

Chapman (1991) estimated that municipal waste incinerator bottom ash could be processed at 45 tonne/day (50 ton/day) in an electric melter for approximately $88/tonne ($80/ton) of ash. However, these numbers have not been validated. In early 1996, GTS Duratek Corporation completed construction of a vitrification facility to treat 2.54 ML (0.67 Mgal) of mixed radioactive waste for a fixed price of $13.4 million. Costs include design, construction, operation, radiological health and safety, and decontamination and decommissioning of nine storage tanks. The vitrification process will operate in 1996 to process the wastes and produce approximately 1.1 Mkg (2.4 Mlb) of glass product at a rate of 300 kg/hr (660 lb/hr).

Combustion melter operation and maintenance costs based on US EPA SITE demonstrations of the Vortec Corporation's combustion melter are $39 to $44/tonne ($35 to $40/ton) for contaminated soils (Shearer et al. 1992). The company estimates capital costs for a 23 tonne/day (25 ton/day) unit to treat hazardous waste dusts produced from arc furnaces and smelting processes, fly ash, and soils contaminated with heavy metals would be on the order of $3.5 million to $4 million in 1991 (Hnat et al. 1991). Capital costs included the costs for feed handling equipment, offgas treatment, glass product handling, and process instrumentation and control. Annual operation and maintenance costs were estimated to be $375,000 to $500,000.

No microwave melters are known to be in commercial operation treating hazardous waste. However, the capital costs for a 60 kW microwave generator, wave guide, tuner, and cavity enclosure are estimated to cost approximately $1 million.

3.3.10.2 In Situ Vitrification

The typical cost of ISV is about $440/tonne ($400/ton) of nonradioactive material treated according to Geosafe Corporation literature. The US EPA performed an economic analysis of ISV at a site with conditions similar to those at the Parsons Chemical Works, Inc. Superfund Site where staging was required (US EPA 1995c). The estimated cost of ISV was $470/tonne ($430/ton). This cost is likely higher than would be expected for the typical ISV site where staging of the contaminated soil into separate cells would not be required. The US EPA (1995c) defined ISV costs according to the 12

Table 3.20
Summary of ISV Costs

Cost Category	Total Costs (%)
1. Site and Facility Preparation	2
2. Permitting and Regulatory Requirements	1
3. Equipment	13
4. Start-up and Fixed	17
5. Labor	19
6. Consumables and Supplies	8
7. Utilities	22
8. Effluent Treatment and Disposal	0
9. Residuals and Waste Shipping/Handling	3
10. Analytical Services	2
11. Facility Modifications and Maintenance	11
12. Site Demobilization	2
13. Long-Term Monitoring	NI

NI   not included in cost analysis

Adapted from US EPA 1995c

categories shown in Table 3.20. Each of the cost categories that represents more than 5% of the total costs is discussed below.

- Equipment costs are dominated by amortization of the major ISV equipment costs ($4 million per ISV system).

- Start-up and fixed costs are dominated by insurance, taxes, and contingency costs.

- Labor costs incurred during melting represent about one-half the total labor cost.

- Electrodes represent the highest consumables cost.

- The cost of electricity is 99% of the utilities cost and about 20% of the total treatment cost.

- Maintenance and replacement of hood panels are the most expensive facility modification and maintenance activities. (Note that hood panel replacement was a requirement at the Parsons site due to conditions there; however, the need has been much less at other sites). Typically, only a few hood panels require replacement per job at a cost of several hundred dollars each.

3.3.11 Design Validation

3.3.11.1 Ex-Situ Melters

Any and all of the ex-situ melters might be applicable to a specific waste stream. Further complicating the decision-making process is the fact that many options exist for each melter type. Therefore, it is important that an objective evaluation method be developed for selecting a system. An assessment could be done as part of a vendor proposal review or as a means to identify and then procure a system. All significant technical, regulatory, and institutional requirements should be identified and assigned weighting factors. The requirements should be further divided into two groups; those which are absolute and the requirements that can be met to varying degrees. Experts that cover the field of application can then be asked to assess each requirement and provide a score to indicate how well each technology or vendor met the requirement. A tally of the scores would indicate which system best meets the requirements of the application. A broad base of technical expertise exists within the personnel of government, industry, and engineering consulting firms who could perform these peer review activities.

3.3.11.2 In Situ Vitrification

The number of individuals qualified to contribute effectively to value engineering and peer review of ISV options is relatively limited. Most of the experts in ISV technology are employees of Battelle Pacific Northwest National Laboratory and Geosafe Corporation. This technology has been applied by Battelle at Oak Ridge National Laboratory and Idaho National Engineering Laboratory. The greatest opportunity for increasing the robustness of the technology is in developing methods for increasing the depth and shape of the melt. Geosafe Corporation may offer certifications and guarantees that ISV will accomplish design objectives at a price that is commensurate with the risks.

3.3.12 Permitting Requirements

3.3.12.1 Ex-Situ Melters

General permitting requirements are well described in Section 3.2.12. Additional regulatory issues associated with ex-situ melters pertain to the treatment and discharge of secondary waste streams. Federal, state, county, and local requirements must be met for the degree of offgas and wastewater cleanup prior to discharge to the environment or municipal treatment system. In many cases, treatment technology must use or be equivalent to BDAT identified by the US EPA to gain regulatory approval. Additionally, process performance data will be required. Data obtained from demonstration tests performed in pilot-scale equipment should be sufficient to gain approval to construct the process and reach agreement on the testing required for the plant once constructed. Work plans can then be prepared detailing the testing to be performed to validate that the pilot-plant data can be duplicated by the full-scale plant. Following plant testing, permitting of the plant would occur and commercial operations could begin.

3.3.12.2 In Situ Vitrification

The application of ISV to hazardous wastes requires permits or approvals in accordance with federal, state, and local requirements. For example, a TSCA Demonstration Permit was required for demonstrating ISV at a PCB-contaminated private site in US EPA Region X. Geosafe now has a national TSCA permit for treatment of PCB-contaminated media anywhere in the U.S. Federal and state permits required at the Parsons Site included a National Pollutant Discharge Elimination System (NPDES) permit to discharge diverted groundwater to a nearby waterway. A Michigan State air permit was also required (US EPA 1994a). The US EPA and state-approved work plans constitute permits for projects governed by CERCLA/Superfund Amendments and Reauthorization Act (SARA) and RCRA/Hazardous and Solid Waste Amendments (HSWA) regulations.

A hazardous waste treatment permit is required when using ISV to process newly-generated industrial wastes. The local air quality control region may also restrict the quantities of air emissions, as well as the types of equipment and fuels that may be used. Local agencies commonly require an excavation permit when the waste is to be dug up and staged prior to ISV. Other local permits may be required to ensure safe operation of a treatment facility and to control discharges of water to a sanitary sewer. Local permits at the Parsons Site were granted by the Department of Building and Safety and by the local fire department (US EPA 1995c). A permit to transport ISV systems across state lines is required because one of the mobile units is overweight.

3.3.13 Performance Measures

3.3.13.1 Ex-Situ Melters

Production rate, reliability, and product quality are the main performance measures. Steady-state production rate is the single most important measure of process performance. This measure determines the economics of the system since it affects energy efficiency, materials consumption, labor costs, and capital cost recovery. The production rate must meet original design expectations and should be maximized to the extent possible. Reliability refers to the on-line factor of the system and the replacement or repair frequency of the subcomponents. As the operating time approaches and is maintained near 100%, the cost per kilogram of waste treated is reduced. Minimizing repair and replacement frequency improves the on-line factor and further reduces operating and maintenance costs. Product quality can be achieved through proper operation of the process coupled to proper feed stream characterization and batching with any glass or chemical additives. Product characterization to determine chemical durability and composition analysis should be performed at a frequency determined necessary for maintaining the processes within required operating parameters.

3.3.13.2 In Situ Vitrification

Implementability, effectiveness, and cost of the ISV process should be evaluated to support decisions regarding its application at a specific site. Key parameters required to assess implementability include:

- mobilization and demobilization time;

- total operating efficiency;

- equipment corrosion and failure rates;

- power and chemical consumption, and;

- rate of secondary waste generation.

Key parameters required to measure the effectiveness of the application include:

- thermocouple temperature readings and geophysical measurements, where applicable, to verify that the target volume has been treated;

- core sample analysis to establish pre-ISV conditions and to verify vitrified product quality;

- temperature and pressure drop data in the hood and offgas treatment train; and

- analyses of samples collected at the offgas treatment system stack before, during, and following operations.

The latter two parametric sets are required to measure effectiveness of the offgas treatment system and compliance with permit conditions.

3.3.14 Design Checklist

3.3.14.1 Ex-Situ Melters

The following key factors should be confirmed to ensure that the design has been properly developed. Depending on the melter type being considered, some of the factors may not be applicable.

1. Sufficiently characterize the waste stream to determine level of homogeneity, corrosiveness, major constituent compositions, primary offgas elements, metals content, and organic content.

2. Define the type of waste pretreatment, such as blending, crushing, sifting, sorting, and/or dewatering required.

3. Determine if any major or minor waste components will be lost from the melter during processing, such as mercury, or have significant volatility, such as lead, cadmium, and halides.

4. Define the required processing period or rate required.

5. Determine the final product property requirements and if there are any anticipated product uses.

6. Identify operating permit requirements.

7. Conduct laboratory studies of waste samples to determine how the waste should be pretreated or if glass or chemical additives need to be added to the waste to achieve an acceptable product.

8. Ascertain if the glass will have acceptable conductivity and viscosity properties at the planned operating temperatures.

9. Determine if the glass is compatible with the materials of construction to be used in the melter.

10. Determine if a secondary combustion chamber is required to treat organics, if present.

11. Configure the offgas system to maximize the recycle of secondary wastes back to the melter without allowing the buildup of any constituents not compatible with vitrification.

12. Determine if the melter is compatible with the required mode of operation, e.g., continuous, intermittent, or batch.

3.3.14.2 In Situ Vitrification

The design phase must determine that conditions are acceptable for vitrification in each melt setting or staged cell.

1. Determine if the waste medium requires compaction to eliminate voids and to destroy the integrity of sealed containers of liquids.

2. Ascertain if the waste medium contains at least 1.4% alkaline oxides and if sufficient silica is present to form a durable glass.

3. Determine the organic content of the waste medium. It should be less than 10% by weight.

4. Determine the elemental metals content of the waste medium. It should be less than 37% by weight.

5. Determine the fraction of inorganic debris in the waste medium. It should be less than 50% by weight.

6. If the site is below the water table, determine if the rate of recharge is acceptable or can be controlled.

7. Determine the depth of material to be vitrified. It should be less than 6.1 m (20 ft).

8. Determine the content of silt and non-swelling clay in the site soils. These fractions should be small enough to ensure safe release of water vapor.

9. Determine if soils at the site can support the crane required for moving the offgas hood.

10. Determine if sufficient space is available at the site to set up and move all required ISV equipment.

11. Employ site modeling to establish appropriate site arrangement, melting characteristics, electrode spacing, and oxide composition of the vitrified medium.

12. Estimate the composition of gases released from the melt, or measure them using treatability testing.

13. Establish offgas emissions limits.

14. Size the offgas blower to contain the air leakage and all gases released from the melt and provide a negative pressure of 1.3 cm (0.5 in.) water in the offgas hood. The blower should be controllable so that these objectives can be satisfied.

15. Design the offgas treatment system to meet emissions limits at the peak offgas flow rates and provide adequate public health protection.

16. Design the site to ensure adequate run-on/runoff control and safe worker ingress and egress.

IMPLEMENTATION AND OPERATION

The information in preceding chapters constitutes the basis for selecting appropriate technologies and initiating engineering design. Each of the technologies covered in this monograph has matured to the point of potential or actual commercial implementation. Current implementation status, as well as issues associated with process startup and full-scale operation are reviewed in this chapter. Operational aspects of monitoring and QA/QC are also covered. Each of the major categories of Stabilization/Solidification (S/S) technologies are discussed in turn; first, Aqueous S/S, followed by Polymer S/S, and Vitrification with ex-situ and in situ applications of each.

4.1 Aqueous Stabilization/Solidification

4.1.1 Implementation

Aqueous S/S processes have been implemented at thousands of remedial sites around the world. Phosphates and organo-clays have recently been used for stabilization alone or for cementitious S/S at a number of sites, and rubber particulate has been used in at least one TSDF, but not yet in a full-scale commercial operation. Cement-slag processes have occasionally been used commercially for chromium reduction; most of these uses have been in proprietary systems and formulations. The ProFix™ process has been used as a combined fixation agent and filter aid. In situ S/S, using auger systems, is now a standard commercial operation, with successful projects numbering in the dozens.

4.1.2 Start-up Procedures

For most remediation projects, startup of innovative aqueous S/S processes follows the same procedures as conventional S/S processes. Equipment is installed and tested, component-by-component, for operation. Test runs may be done at startup, but more often full-scale treatment is started with any problems worked out along the way. This is possible for ex-situ processes because of the vast experience available and the operational similarity of all the ex-situ systems.

Auger-type in situ treatment usually requires a pilot-scale test because the quality and uniformity of the waste to be treated is seldom known with any degree of certainty. Frequently, it is necessary to modify the working "tool" to achieve proper drilling rates and uniformity of mixing. Modification of injection arrangements might also be required. The presence of debris can be the major working impediment in the systems. These modifications and adjustments are done on an experiential basis by operators skilled in this technology and are generally held as proprietary information.

4.1.3 Operations Practices

Operations practices in innovative aqueous S/S projects are taken directly from long-established practices in conventional S/S, since the only operational difference is the use of different reagents. The one exception is auger-type in situ projects. Even these are based on decades of experience in constructing foundations and cutoff walls. Therefore, the reader is referred to standard handbooks and case histories of conventional S/S operations for more details.

4.1.4 Operations Monitoring

Table 3.8 lists some of the test methods used to monitor process parameters in an aqueous S/S system. Continuous monitoring is required for feed rates of waste, reagents, and water. Mixer speed is usually fixed and requires only periodic monitoring to ensure that excessive buildup does not occur in the mixer. Overall system parameters include waste input and output rates and tests of incoming waste and treated product for the appropriate parameters (Table 3.8), as well as solids content.

Table 4.1
Summary of Standard Methods and Procedures

Parameter	Units	Method	Method Title	Method Type	Reference
Total Solids/Moisture Content	%	EPA 160.1; ASTM D2216-80	Water (Moisture) Content of Soil, Rock, and Soil-Aggregate Mixtures	Drying	ASTM
Bulk Density	g/cm^3; lb/ft^3		Apparent Bulk Density Measurement	Gravimetric/Volumetric	
Free Liquid	pass/fail	EPA SW846, Method 9095	Paint Filter Test	Filter	SW846
Penetration Resistance	psi	ASTM C-403-80	Penetration Resistance	Pressure	ASTM
Unconfined Compressive Strength	psi	ASTM D-1633-84; D-2166-85; C-109-86	Compressive Strength of Molded Soil-Cement Cylinders/of Cohesive Soil	Compressive Strength	ASTM
Permeability	cm/s	EPA SW846, Method 9100	Permeability	Hydraulic Conductivity	SW846
pH	pH units	EPA SW846	pH	pH	SW846
Temperature Rise	°C~		Temperature Rise for Pozzolanic Stabilization Agents	Temperature	
Binder and Additive Mix Ratios	ratio		Stabilization Formulation Method	Gravimetric	
Color Change	color		Stabilization Formulation Method	Visual	
Total Constituent Analyses	mg/kg	EPA, SW846 Methods, other methods as appropriate (see Table 2.1)	Analytical Methods	Total Analysis	SW846
Leachability					
TCLP	mg/L	SW846-1311, followed by appropriate analytical method	Toxicity Characteristic Leaching Procedure	Leaching	40 CFR Part 268, Appendix I

4.1.5 Quality Assurance/Quality Control

Product QA/QC in wet stabilization involves any or all of the parameters listed in Table 4.1 along with brief descriptions and references. Data quality objectives are determined by standard US EPA practices which apply to aqueous S/S and need not be discussed further here. Methods for sampling, storage, handling, chain-of-custody, and analysis/testing are also standard US EPA and ASTM methods and practices. A good general description of QA/QC requirements is given in the US EPA handbook, *Preparing Perfect Project Plans* (US EPA 1989b).

No overall performance standards exist for remedial projects because each project's requirements are set by site-specific regulatory, environmental, operational, and future use considerations. However, certain benchmarks are fairly common in aqueous S/S; a listing of these is given in Table 4.2. An example of standards for one project, as well as the performance achieved in the project, are given in Table 4.3.

Table 4.2
Typical Aqueous S/S Performance Benchmarks

Parameter	Performance Standard
Unconfined Compressive Strength	50 psi
Permeability	$1 \cdot 10^{-6}$ cm/sec
TCLP Leachability	RCRA Toxicity Characteristic Standard for Metals
Modified ANS 16.1 Leachability	Leachability Index $= \geq 12$

*These parameters can vary greatly depending on the end use for the project site or final disposition of the waste products.

Table 4.3
Performance Standards and Performance
Achieved in an Actual Aqueous S/S Project

Parameter	Standard	Performance Achieved
Lead Concentration in Waste		50,343 mg/kg
Lead Leachability of Raw Waste		EPT-35.8 mg/L; TCLP-355 mg/L
Binder:Soil Ratio		0.4:1, or 40%
Water Addition Ratio		0.23:1, or 23%
Weight Increase		40% without water, 63% with water addition
Volume Increase		37% including water addition, 18% due to chemical additives
Strength (Penetrometer)	2000 psi	> 9,000 psi
Unconfined Compressive Strength	2000 psi	11,720 psi
Permeability	10^{-6} cm/sec	$7 3 \cdot 10^{-6}$ cm/sec
TCLP Leaching	5.0 mg/L	0.62 mg/L for lead
EPT Leaching	5.0 mg/L	0.35 mg/L for lead
Modified TCLP Leaching		< 0.04 mg/L after filtration
Modified ANS 16.1 Leaching (Full-Term Test)[*]	≥ 12	13.8
Diffusion Coefficient	$\leq 10^{-12}$	$< 10^{-14}$ for all intervals except the initial washoff

*Leach Index as defined in ANS 16.1 (ANSI/ANS 1986)

4.2 Polymer Stabilization/Solidification

4.2.1 Implementation

Research, development, testing, and demonstration of polyethylene encapsulation has been supported by the DOE at Brookhaven National Laboratory over the last twelve years and at the Rocky Flats Plant over the last four years. Several private sector vendors have expressed interest in commercializing the technology. For example, Brookhaven National Laboratory completed a Cooperative Research and Development Agreement (CRADA) with

MMT of Tennessee and negotiated agreements with several other companies. Recently, DOE contracted with Envirocare of Utah, Inc., to commercialize polyethylene macroencapsulation. Their contract calls for processing up to 226,800 kg (500,000 lb) of DOE mixed waste lead in exchange for development incentives (see Section 5.2.2). Brookhaven National Laboratory has provided technical assistance to Envirocare in their implementation effort.

Sulfur polymer encapsulation development has also been supported by DOE at Brookhaven National Laboratory for about 11 years. A process application patent for the technology is pending. The Idaho National Engineering Laboratory has conducted bench- and pilot-scale testing of sulfur polymer encapsulation over the past five years. More recently, the process has received attention at Oak Ridge National Laboratory and other national laboratories. Currently, Brookhaven National Laboratory is engaged in a CRADA with Scientific Ecology Group, Oak Ridge, Tennessee, for the commercial demonstration and application of the SPC encapsulation process and other companies have signed Research and Development license agreements with Brookhaven National Laboratory. The Scientific Ecology Group has used the process to stabilize and solidify mixed waste contaminated incinerator fly ash from their commercially-licensed volume reduction facility.

In situ polymer S/S has been under development at Brookhaven National Laboratory for about four years. Recently, collaborative projects to develop, test, and demonstrate this technology have been launched with Sandia National Laboratory, Idaho National Engineering Laboratory, and Hanford. A small field test of in situ S/S was recently demonstrated at Idaho National Engineering Laboratory (Loomis, Thompson, and Heiser 1995). An acrylic polymer was used to stabilize a pit of buried waste that consisted of drums, metals, rags, and paper. The polymer was delivered via jet grouting at ~34.5 MPa (5,000 psi). Excavation of the demonstration site proved the waste was successfully stabilized. Examination of cored samples indicated homogeneous mixing, absence of voids, and a monolithic solid structure. Additional testing work to characterize cored samples is planned. Several companies have expressed interest in commercializing in situ polymer S/S, but no companies have implemented this technology to date.

4.2.2 Startup Procedures

As with any new treatment technology, startup of a full-scale polyethylene encapsulation process requires integration and "cold" (i.e., non-radioactive and nonhazardous) testing of all system components (e.g., pretreatment, materials handling, metering, extrusion, process control). Non-radioactive and nonhazardous surrogate waste materials should be used in testing individual components and overall system operation.

Following installation of dryer pretreatment equipment, all liquid lines (especially high pressure steam) should be examined for leaks. Initial dryer operational checks should be conducted using clean water. Pre-operational checkout of the material handling system should include inspection of vacuum pumps, lines, and gate valves prior to transferring actual materials. Metering equipment should be calibrated on a volume and/or mass basis. Extruder installation requires equipment leveling and bore-scoping of the barrel for proper alignment. Following installation of electrical wiring, shaft rotation should be checked to ensure proper motor polarity. The protective corrosion coating applied to the barrel and screw should be removed with a light solvent. Initial extruder operation should be conducted using 100% polyethylene until troubleshooting is completed. Process control equipment should be tested and optimized to maintain proper feeding rates within de-fined operating parameters. On-line monitoring equipment should be cali-brated at several known waste loadings and various process rates within prescribed operating specifications. Following installation and pre-opera-tional checks of each subsystem, confirmation testing of the integrated pro-cess is recommended.

Sulfur polymer cement encapsulation equipment should be thoroughly tested using "cold" surrogate materials prior to hot testing. All steam, hot water, or oil circulation lines should be checked to ensure leak-proof opera-tion. Thermocouples should be calibrated and inspected for proper function-ing. Successful operation of vacuum and venting systems, as well as process valves, should be verified. Mixer interlocks and other safety-related systems should be checked.

For in situ polymer S/S, the selected resin should be compatible with the waste types and geochemistry of the site. Selection can be accomplished by consulting polymer manufacturers. The shelf lives of the monomer, cata-lyst, and promoter should be noted. The manufacturer can provide data on shelf lives for the resins and for the pre-promoted and pre-catalyzed mixes. If the shelf life is long enough, the resin mixes can be delivered to the site pre-mixed by the manufacturer. If the shelf life is only days for the pre-mix, then the catalyst and/or the promoter might have to be mixed on-site.

In most cases, it is wise to hire experienced contractors and use their equipment. Select a contractor who has previous experience in polymer grouting and, in particular, two-part delivery using dual wall drill stems. To protect the equipment, the polymerization reaction should only occur after the grout leaves the drill stem. For this reason, the dual wall drill stem is used to deliver the resin in two parts: (1) one part containing the resin charged with catalyst and (2) the other containing the resin pre-mixed with the promoter. Mixing and pumping of the resins can be done with conventional equipment, so long as the compatibility of the parts (e.g., seals, diaphragms, etc.) is ensured. The resin manufacturer should have a full listing of compatible materials and pumps/mixers.

4.2.3 Operations Practices

4.2.3.1 Polyethylene Encapsulation

Operation of the polyethylene process is designed to minimize on-line operator adjustments. Overall process rate depends on screw speed, which is directly controlled by the operator by adjusting the speed dashpot. While observing extruder output rate, the operator gradually increases screw speed until the desired output is attained. Feeder controls are automatically adjusted at the pre-set ratios to match the demand. Alternatively, this operation can be tied into the process control system by means of a solid-state speed controller and appropriate modifications to the process control software. In this way, process rate can be pre-selected for even greater automated control.

As discussed in Section 3.2.4, a specific temperature profile can be set by adjusting barrel and die zone temperature controllers. Optimal temperature conditions will depend on the type of polymer (melt temperature, melt viscosity) or polymer blend (including the presence of recycled polymers), waste type, and waste loading. In general, temperatures are maintained at minimal levels to ensure adequate melting while minimizing offgas potential. Higher temperatures can be used to reduce melt viscosity, enhancing mixing and optimizing waste loading potential. Usually, a gradually "stepped" temperature profile is used in which the temperature in each zone along the length of the extruder is increased from the previous zone. This ensures gradual heating of the polymer throughout the barrel. The temperature is kept lower in the feed zone to maintain the frictional forces needed to convey the mixture forward.

Process startup should always be conducted without waste (100% polymer) and at moderate speed (e.g., 25% of maximum speed) until steady-state operation is confirmed. After the process is operating smoothly, the waste feed is introduced gradually (e.g., increments of 10% by weight) while the operator confirms proper feeding and continued steady-state flow at the extruder feed throat, inspection port, and output. Changing the waste loading from 0% (by weight) directly to the final waste loading (e.g., 60% by weight) can cause the waste feed controller to overshoot the target feed rate, resulting in a temporary condition in which maximum processing limits are exceeded. If this occurs, the extruder can become plugged with waste, necessitating system shutdown and purging. Once the target waste loading is achieved, the screw speed is gradually increased to maximum output (determined empirically based on the type of waste and polymer, waste loading, etc.).

Routine maintenance is conducted according to equipment manufacturer's recommendations. Depending on the chemical and physical properties of the wastes being processed, periodic inspection of equipment directly in contact with waste (dryer vessel, hoppers, extruder barrel, and screw) should be conducted to monitor corrosion and/or excessive wear. Extruder screws are removed using a hydraulic jack to push the screw from the barrel, while supporting and guiding the screw with an overhead crane.

Standard OSHA safety practices should be followed for operation of equipment at elevated temperatures and pressures. The extruder control module can be equipped with temperature, pressure, and current load alarms and automatic shutdown capability in the event of a process excursion. In the rare event of an over-pressurization of the extruder barrel that goes uncorrected, a rupture disk provides emergency pressure relief. As in any process for hazardous or radioactive materials, appropriate ventilation and fire protection systems are recommended.

4.2.3.2 Sulfur Polymer Cement Encapsulation

Sulfur polymer cement encapsulation processing is a batch process and, as such, is somewhat forgiving in terms of the actual steps of operation. However, depending on the type and nature of the waste to be processed, order of addition for the waste, binder, and additives, as well as the form of SPC charged to the mixture, can affect processing success. For example, for wastes considered difficult to handle and mix, preheating the waste mixture and pre-melting the SPC before combining and mixing, expedite the mixing process. Slowly adding waste to molten SPC or vice-versa might help avoid a rapid decrease in the temperature of the mix and premature "freezing" of the molten mixture. Small amounts of sulfur that volatilize during processing can cool rapidly in ventilation lines, so care should be taken to inspect and clean lines as necessary.

4.2.3.3 In Situ Polymer Stabilization/Solidification

Grout injection holes should be carefully planned prior to operation to ensure complete coverage of the contaminated site. Estimates of grout requirements must be made based on existing site void volume. Since ratios of monomer to catalyst and monomer to promoter are critical to successful S/S, flow rates should be calibrated. This can be accomplished using other fluids of similar viscosity/density (e.g., water). Test drilling under similar soil conditions should be conducted to optimize drilling parameters, including vertical step (withdrawal) rate, rotations/step, jet orifice diameter, and polymer grout pressures. Optimizing these parameters maximizes grout coverage and minimizes the volume of spoils (rejected grout/soil mixture). The grout injection sequence should be planned so as to minimize cross-talk (breakthrough of one hole to adjacent areas) and placement and setup time. Typically, injection holes are drilled in an alternating sequence. On completion of drilling operations, all resin delivery lines/pumps should be purged and flushed with appropriate solvents (e.g., detergent solutions and/or diesel fuel).

4.2.4 Operations Monitoring

As discussed in Section 3.2.7, process parameters, including heating and die zone temperature profiles, melt temperature, melt pressure, vacuum, screw speed, and current load, are continually monitored through the extruder control module. Overall system parameters including output rate, polyethylene and waste feed rates, and waste/binder ratios, are monitored by a computerized process control system. Actual waste loading is monitored by the on-line Transient Infrared Spectroscopy (TIRS) monitor.

The temperatures of the mixture and heat transfer medium are monitored constantly. In addition, the operator should track vacuum level, mixing speed, and motor load. A closed-circuit video camera allows the operator to observe the mixing vessel at all times.

If in situ S/S is conducted in an enclosed area (e.g., tent), monitoring of potentially hazardous constituents and/or nuisance odors emanating from either the soil contaminants or the grout materials might be required. Injection flow meters should be checked on a regular basis to verify proper flow and correct mixing ratios. Flow meter measurements can be checked by monitoring the decrease in storage tank volumes over time.

4.2.5 Quality Assurance/Quality Control

For any S/S process, QA/QC activities are conducted for the process and the final waste form product. To ensure the polyethylene encapsulation system is operating within previously-defined processing specifications leading to successful final waste forms, each of the key process parameters described in Section 4.2.4 are monitored. Waste loading, one of the most critical parameters, is monitored in real-time to provide an on-line evaluation of QA/QC. Since the polyethylene encapsulation process is not affected by changes in waste chemistry, variations in the waste chemical composition do not impact processing QA/QC. Melt viscosity can be confirmed for known polymers or determined for polymer blends using a melt indexer and appropriate ASTM standard test methods.

Final waste form QA/QC can be monitored by evaluating grab samples taken periodically from the output stream and core samples taken from final waste forms at pre-determined intervals. Grab samples confirm output rates and homogeneous mixing (by measuring product density). The latter is measured for small grab samples (≤ 5 g) using a pycnometer. Final waste form properties discussed in Section 3.2.13 should be confirmed on a regular basis by conducting the appropriate standard tests (ASTM, Nuclear Regulatory Commission (NRC), US EPA, American Nuclear Society, etc.) on cored, full-scale samples and comparing the data with previously-generated, bench-scale data.

Process QA/QC for SPC involves monitoring key parameters (e.g., temperature, mixing speed) as discussed in Section 4.2.4. Periodic observation of the mixing process confirms successful operation. Material specifications confirming that SPC meets applicable ASTM standards should be provided by the manufacturer. Confirmation of final waste form properties should be conducted on a regular basis by taking a grab sample of each batch processed. Approximately two representative 1 L samples of the molten SPC/waste mixture should be collected and saved for archival purposes. One sample should be used for mandatory QA/QC testing of product density. This information verifies proper waste loading and homogeneous mixing. The second set of grab samples can be re-melted and cast into suitable bench-scale test specimens, as necessary, for periodic confirmation of waste form performance. Testing should include compressive strength and applicable leachability tests.

For in situ polymer S/S, QA/QC confirmation involves monitoring and recording injection pressures and flow rates, as well as drilling speed and depth. Samples of the catalyzed and promoted monomers should be taken for archival purposes. Following completion of one "hole" or section, the soil/polymer mix should be sampled for archiving prior to solidification. Once in situ polymer S/S is completed, representative samples of the solidified soil can be cored for confirmation and performance testing, including compressive strength and leachability. Depth of penetration can be monitored directly for deep soil mixing and jet grouting based on how deeply the auger or drill is extended. For permeation grouting, indirect confirmation techniques, such as ground-penetrating radar or acoustical sensors, are required.

4.3 Vitrification

4.3.1 Implementation

4.3.1.1 Ex-Situ Melters

Vitrification technology has been extensively developed and demonstrated by the US EPA, DoD, and DOE. This has resulted in the establishment of a significant technology application and adaptation expertise within many of the government research laboratories. Private companies, such as Geosafe Corporation and Vortec Corporation are aggressively pursuing contracts to apply government-developed and government-funded technologies. Still other technologies, such as the Babcock and Wilcox cyclone furnace, have been developed solely by industry without government support. As a result, ex-situ vitrification technologies are available from many different sources.

As a generalization, most any site can be remediated using any of the ex-situ melter technologies, with the possible exception of microwave melting. Howeve, there are certain wastes for which melter technologies are best suited. It is important that the engineer responsible for implementing the ex-situ melter technology understands the relative strengths and weaknesses of the technologies considered. Most vitrification technologies are not offered as turnkey operations — nor are they offered by companies that would provide operations as a part of the contract — rather, they are procured using the conventional design-build-operate method. If a turnkey operation is required, it should be expected that many melter vendors would team with consulting engineering companies, or remediation companies that could provide the necessary skills. It is also important to remember that vitrification technologies have not yet been demonstrated for many of the potential applications. This dearth of data limits vendor claims regarding process effectiveness and leaves uncertainty about design, construction, and operation costs. As a result, the construction contract bid and award process should require demonstration tests of the waste stream at the vendor's facility to verify process and product claims by the vendor.

The preferred contract terms would provide payment to the remediation company based on treating all the waste within a specified period of time. Such a lump sum contract has been awarded to GTS Duratek by the DOE to treat radioactive solid and liquid waste at the Savannah River Site in Aiken, South Carolina. In this case, the waste volume, composition, and physical and chemical properties are well-documented. GTS Duratek is in the process of designing and constructing the vitrification facility, including tank sluicing, feed preparation, and offgas treatment processes.

If the waste is poorly characterized or heterogeneous, it might be necessary to negotiate a contract in which unit price payment is based on the volume or mass of waste treated. The DOE is in the process of attempting to apply this strategy at the Hanford Site in Richland, Washington. In this case, a down-selection process is used to select vendors to specify, design, construct, and operate vitrification facilities to treat low-level and high-level radioactive tank waste. These vendors will not be paid until they begin to process the waste and produce a glass product.

4.3.1.2 In Situ Vitrification

The rights to ISV technology are owned by the U.S. government under the DOE and by Geosafe Corporation, a company formed by the inventor (Battelle Pacific Northwest National Laboratory) to commercialize the technology. Battelle Pacific Northwest National Laboratory is continuing to develop the technology for applications at federal sites. To date, Battelle Pacific Northwest National Laboratory has assisted in ISV demonstrations at several DOE sites, including Hanford (Luey et al. 1992), Oak Ridge National Laboratory (Spalding et al. 1992), and Idaho National Engineering Laboratory (Callow et al. 1991). Battelle continues to offer ISV treatability testing and demonstration services at federal facilities.

Currently, Geosafe Corporation is the exclusive worldwide licensee for ISV in applications outside the federal government. In 1995, Geosafe licensed ISV-Japan Limited for applications of ISV within Japan. ISV-Japan is a joint venture between Geosafe and five leading Japanese companies (Thompson, McElroy, and Timmerman 1995).

To use ISV to remediate a commercial waste site, it is necessary to either obtain a license or contract the services of a licensee. The technology is available for use by anyone who plans to remediate a federal waste site. The ISV services of both Battelle Pacific Northwest National Laboratory and Geosafe Corporation are available through a variety of contractual mechanisms. The testing and demonstration services provided by Battelle Pacific Northwest National Laboratory are probably best obtained through a cost-plus-fixed-fee/unit price contract due to the high level of uncertainties associated with this type of work. Geosafe Corporation can enter into full-service, turnkey, and design-build-operate contracts under virtually any type of compensation method.

4.3.2 Start-up Procedures

4.3.2.1 Ex-Situ Melters

Electric Melters. Melters containing castable and/or fused-cast refractories are started gradually to allow the castable refractory to cure and the fused-cast refractory to heat up without fracturing. Prior to this, a glass cullet or frit should be placed in the melter furnace to serve as the start-up charge. The heat source is supplied by electric or gas heaters suspended in the furnace.

Refractory manufacturers publish temperature ramp and hold point data for each refractory. Once the furnace refractories have been cured, the electrical continuity between the electric wire connections penetrating the melter shell and the melter shell should be measured to ensure that the electric isolation designed into the system is intact. Once completed, the rest of the bake-out cycle can be completed. During the bake-out period, it is wise to circulate a small amount of air through the furnace plenum space to facilitate removing the water vapor released from the castable refractory.

The furnace temperature is steadily increased until the minimum temperature required for electrical conduction through the glass is achieved. With the electrode transformer set at the maximum voltage, a low amperage current is conducted through the glass. Once conduction is achieved, the amperage rate should be increased very slowly. Otherwise, the glass will be overheated along a very small cross-section and furnace damage could result. As the voltage drops between the electrodes, the transformer taps should be adjusted to keep the voltage within the upper quartile of the tap settings; this minimizes phase instabilities in the electrical supply.

Once the furnace is idling near its operating temperature, electrical isolation between the electrical connections and the shell should again be checked. All bolts and fasteners securing the various components should be checked and tightened to compensate for the thermal expansion that occurred during the startup. At this time, the "air tightness" of the melter can be checked to determine the amount of in-leakage that occurs when the offgas system is in service. If excessive, operators should check all gaskets and tighten all flanges and pipe connections. Some in-leakage can be tolerated, but it should not be excessive. At this time, offgas process equipment, e.g., pumps, valves, and filter blowback systems, and instrumentation, e.g., flow, temperature, liquid level, and pressure monitors, should be checked for proper functioning and operation. The following start-up items should be completed next:

- add remaining glass charge to bring melter level to its full inventory;

- determine that all cooling jackets and channels are performing properly;

- observe electrode power, voltage, and amperage to verify proper operation;

- verify controller performance by entering set point changes and observing responses;

- test glass discharge system to verify leak tightness and glass pour control;

- connect feeding system to the melter and prepare for test feeding;

- feed melter at 25% capacity to verify feed and offgas system performance is within expected operating ranges; and

- gradually increase feeding rate to achieve maximum designed surface coverage.

Be aware that large electric melters might require one or more days to achieve steady-state mixing patterns. Therefore, do not increase the feed rate more than a few percent per hour until the maximum coverage is reached. This can take a matter of days. If the glass is actively mixed, this cautionary note does not apply. At the end of the acceptance testing phase, the glass tank should be probed to verify that there are no regions in which the glass "feels" significantly more viscous or that crystalline phases are accumulating. If crystalline phases are present, the probe will feel like it is pushing through a sandy or silty layer. If such viscous or crystalline regions are observed, the melter insulation is either too low in that area, or, if multiple electrode pairs are used, the electrode power was not sufficient to properly heat the area.

The glass product composition should be analyzed for the entire duration of the acceptance test to verify proper mixing of the glass tank. If no agitation is used, the change in glass composition should approximate the theoretical well-mixed tank model described as:

$$C/C_o = 1 - e^{-(t \cdot v/V)} \qquad (4.1)$$

where: C = the concentration of tracer in the discharged glass;
C_o = the concentration of tracer in the feed entering the melter;
t = time;
v = the glass feed rate to the melter; and
V = the melter tank volume.

Analysis of the measured data will indicate whether there are significant stagnant zones within the melter or if poor mixing results in a plug flow effect.

Once the melter has gone through initial startup and checkout, it can be idled at a temperature less than the operating temperature but above the liquidus temperature of the glass. Prior to the start of routine feeding, the glass temperature should be increased to the operating temperature; instrumentation readings should be checked to verify that operating parameters are within limits and that the instrumentation is functioning. The offgas ventilation should be started and adjusted to achieve the proper negative pressure within the melter plenum, and the glass discharge area temperature should be increased to the operating temperature. Then, waste feeding can be initiated.

Combustion Melters. A benefit of a combustion melter is the fact that they can be idled in a cold state and attain operating temperatures within a matter of hours. Depending on the furnace design, initial startup following installation includes the curing or drying of the refractory lining. This would be similar to the process described for electric melters. Initial startup will cover the applicable items described above, e.g., leak tightness, secondary heaters, glass discharge, ventilation control, etc. Additionally, the combustion melter fuel system and combustion air supply and preheater must be inspected to verify leak tightness. Optimal efficient operation of a combustion melter relies on "tuning" the fuel, primary, and secondary air parameters. If the furnace unit or the waste stream are significantly different than past experience, testing is required to optimize the set points. Routine operation is straightforward and follows the start-up sequence described for electric melters.

Induction Melters. Induction melters follow a simplified version of the electric melter startup. Lacking refractory in the melter tank simplifies the initial startup. A glass charge is placed in the melter, and start-up heaters heat the charge. At a certain temperature, the glass couples to the induced current and the start-up heaters can be removed. Once operating temperature is achieved, the power should be turned off to allow close inspection of the induction coils and cooling circuits to verify that none have warped and that water is flowing through each one. Following this step, the glass discharge system can be tested to verify proper operation. Several tank volumes of test feed material should then be processed to verify proper tank operation as was described for the electric melter. Routine operation also duplicates the electric melter start-up sequence.

Microwave Melters. No special start-up activities are required for microwave melters. Once construction is complete and electrical and control systems have been checked, the microwave generator and furnace can be started. At startup, staff should inspect the microwave generator, wave guide, tuner, and furnace for any microwave leakage with hand-held detectors. Initial process tests would then be conducted to determine optimal power input rates and feed rates.

4.3.2.2 In Situ Vitrification

Startup of the ISV process begins with insertion of four graphite electrodes about one foot into the ground in a square array at the required spacing. Shallow trenches are dug between each of the electrodes. The trenches are subsequently filled with a mixture of glass frit and graphite powder to provide highly conductive current paths. The offgas hood is installed over the electrodes and then sealed to both the ground and the electrodes to limit ingress of air when the hood is operated under negative pressure. The hood is connected to the offgas treatment system, and all electrical, instrumentation, and control circuits are completed.

Operational and management oversight checklists are completed before power is applied to the system. Pre-start-up checks include bumping motors to ensure operation and proper rotational direction and calibrating instruments where necessary. The offgas system is started first and, when proper performance has been verified, power is applied to the electrodes. Electrode power is gradually increased to the limits of the equipment and the regulatory permit until steady-state power input is reached after about 24 hours.

4.3.3 Operation Practices

4.3.3.1 Ex-Situ Melters

Electric Melters. At steady-state operation, electric melters are extremely stable and easy to operate. Generally, the glass temperature is maintained within an operating range by a controller that automatically adjusts the electrode current based on a temperature feedback signal. The temperature can be measured directly with thermocouples or a pyrometer, or the temperature can be inferred by measuring refractory wall temperatures. The latter two cases are attractive because they preclude the need for maintaining a set of thermocouples in a thermowell within the glass. However, sufficient experience should be acquired correlating glass temperature measurements to pyrometer or refractory, and periodic tank probes with a thermocouple should be performed. Alternatively, if the waste stream and resulting glass composition are consistent over time, the electrode power can be controlled by maintaining constant power or current between electrodes; or the calculated resistance between the electrodes can be used as a control parameter. This control measure is possible because glass resistivity is inversely proportional to glass temperature. Therefore, very reliable temperature control can be maintained without intrusive temperature measurements.

Secondary heaters can be controlled using continuous or periodic temperature measurements in the heated zone. Plenum space pressure, feed and offgas line flows and pressures are measured and controlled by standard industrial devices. Level sensors, such as bubblers, have been used to monitor glass level in the melt tank; however, they are prone to rapid erosion and must be frequently replaced.

The primary control parameter for electric melters is the feed rate. The feed rate can be steadily increased until the glass surface is nearly covered by the cold cap — the optimal situation. The melting and spreading properties of the feed stream determine whether optimal coverage can be achieved, and periodic visual observation is the best means to monitor processing conditions as a function of feed rate and coverage. Processing stability can also be tracked by monitoring the plenum space temperature and offgas composition if an offgas product, such as NO_x, is generated.

Process upsets result if overfeeding occurs, cooling water enters the melter, or if ventilation is interrupted. When a melter is overfed, the cold cap can form a rigid bridge between opposite walls of the melter. As feeding continues, the feed accumulates on top of the cold cap. At the same time, the cold cap material contacting the molten pool melts, creating a vapor space between the bridged cap and the glass pool. When this happens, heat transfer to the cold cap drops significantly, and the feed accumulation rate increases. As this occurs, the plenum space temperature drops quickly. If the glass temperature is maintained using temperature or resistance feedback, the electrode power drops in response to the drop in heat transfer to the cold cap. If a constant current or power input is used for electrode power, the glass temperature increases. If a slurry is being fed to the melter and a crack forms in the bridge allowing slurry to pour onto the glass surface, a steam surge occurs which could pressurize the melter. This pressure can blow out gaskets, and a glass surge into the overflow could occur, damaging heaters or plugging the discharge area. In an extreme case, the melter itself could be damaged. Partly to preclude such an occurrence, melters have an emergency or secondary vent line that would open automatically if the plenum pressure increases above a set point. When overfeeding or bridging is observed, the feed stream should be stopped and restarted only after the cold cap has returned to a normal state.

If cooling water were to enter the melter due to a leaking cooling line, steam surges can occur, which disrupt normal operation. Usually the pressure of water supply lines that could fail and leak water into a melter are set at only a few pounds pressure to avoid the situation whereby a water jet could be forced into the glass pool and create a vigorous steam surge. The only response is to stop the water leak and allow the melter to vent until it has boiled off all of the steam.

The loss of offgas ventilation can result from a failed valve, tripped blower, or a plug in the offgas line between the melter and first scrubber. The first two events would happen without warning. If this occurs, melter feeding should stop and the emergency or secondary vent opened to allow the accumulated feed to react and melt into the glass pool. A plugged offgas line can normally be observed by noting changes in the relative pressure in the offgas line. However, significant changes might not be observed until the blockage is nearly complete. When it is apparent that a blockage is occurring, the feed to the melter should be stopped and the plug located and removed.

Combustion Melters. High process rates and short retention times characterize the combustion melter; its operation is straightforward and reliable. The primary control parameters are feed rate, thermal load, primary and secondary combustion air flow rate, and temperature. The process is primarily controlled by steadily increasing the feed, fuel, and combustion air rates until the maximum rate is reached, at which an acceptable glass product is produced. For feed streams containing high concentrations of alkali metals, an upper temperature limit could be based on minimizing vaporization losses. For feed material that is pneumatically conveyed into the furnace, the air pressure and flow rate can be adjusted to increase or decrease the dispersion of the material into the furnace. Process control consists of monitoring furnace chamber and exit temperatures and the glass product consistency. On-line offgas analyses of oxygen, NO_x, and nitrogen also indicate whether the process is operating optimally over time.

No significant upset conditions have been identified for combustion melters. Assuming stable air and fuel supplies and reliable feeding systems, no major events could be considered upset conditions. Feed line blockages are readily observed from feed flow rate measurements. If they do occur, the fuel and combustion air are simply reduced until the feed system is restored. Should either the fuel or combustion air flows become unstable, the system can be quickly shut down and the problem corrected.

Induction Melters. Induction melter systems behave similarly to electric melters. Although induction melters are considerably smaller, their through-put is greater on a glass surface area basis. Therefore, they can change operating behavior more quickly (e.g., cold cap coverage or average bulk glass temperature). As a result, reliable process monitoring and operator surveillance become more important. The temperature of the glass can be measured with thermocouples or pyrometers. Cold cap size should be observed through a view port. Upset conditions are similar to those described for electric melters.

Microwave Melters. As a batch melting process, microwave melters are simple to operate. Melt rate is controlled by the microwave power level and the feed rate into the container. The batch fill rate is controlled based on visual observations, process knowledge, and thermocouple or pyrometer measurements. For maximum processing rates, power levels should be maintained as high as possible while still maintaining even heating throughout the melting zone.

A potential upset is an event described as "thermal runaway" (White, Peterson, and Johnson 1986). Thermal runaway occurs when the local heating rate in the material exceeds the rate at which heat diffuses through the material. These hot spots subsequently absorb more energy than the surrounding material because microwave absorption increases with temperature in most materials. Thermal runaway can be prevented by:

- carefully controlling the microwave power while continuously monitoring the temperature of the material; and

- providing a uniform microwave illumination of the process material.

4.3.3.2 In Situ Vitrification

Performance costs are minimized when the ISV system is operated at the maximum power levels that still conform to all design and permit limits. This speeds the rate of remediation, resulting in lower personnel costs and lower heat losses to the offgas system. Several factors limit the ability to attain maximum power levels, including combustion of pyrolysis products in the offgas hood, which might cause excessive offgas temperatures. In this case, power to the electrodes should be reduced to slow the generation of pyrolysis products.

A high water table can also limit performance by requiring higher energy input to evaporate the excess water. This limitation can be minimized at some sites that are amenable to draining or lowering the water table, and thereby reduce energy demands. Limits imposed on emissions of constituents of concern in the offgas also impede optimal performance if the offgas system is not adequately designed to remove or destroy the concentration of hazardous offgas constituents produced at maximum power levels.

Potential process upsets include the formation of a hot cap within the hood, pressurization of the hood due to excessive gas generation rates, and excessive agitation of the molten glass inside the hood. The appropriate response to each of these upset conditions is to curtail or reduce power to the electrodes. Emergency power cutoff switches ensure rapid response, if necessary. The offgas system is left running in an emergency to ensure continuing treatment of hazardous offgases for an adequate time after power to the electrodes is curtailed. A diesel-powered backup offgas treatment system is used in case of power outage.

The most significant routine ISV maintenance activity is adding sections of new graphite to the top of the graphite electrodes to increase the length of the electrodes as the melt depth increases. Electrode-to-power cable connections and electrode-to-offgas hood seals also require maintenance. Adding electrode sections and other forms of maintenance requiring access to the offgas hood are accomplished after power to the electrodes is cut off.

4.3.4 Operations Monitoring

4.3.4.1 Ex-Situ Melters

All ex-situ melters rely on monitoring and controlling:

- feed rate and distribution;

- energy input; and

- temperature.

If each of these variables is adequately monitored, the process can operate optimally — a state best achieved by directly measuring, rather than inferring by indirect measurement, each variable. Process optimization assumes that the combination of waste, glass, and chemical additives has been carefully determined to optimize (1) reaction and melting of the batch, and (2) glass processing properties, such as viscosity, electrical conductivity (when important), and phase stability. The critical importance of this initial work cannot be too strongly stressed. In addition to these key variables, process parameters that should be routinely measured to allow the process engineer to "see the whole picture" include plenum and offgas system temperatures and pressures, feed, glass product and offgas composition, and energy supply measurements. Key parameters should be displayed and logged continuously in a graphic format to allow historical trending analyses.

4.3.4.2 In Situ Vitrification

Approximately 50 system parameters are monitored during ISV operations. Alarm levels and responses are established for each of the parameters. Visual alarm indicators are color-coded to aid in defining the appropriate response. Trends in parametric data are monitored to aid in fine-tuning electrode power levels. Offgas sampling and analyses are also conducted to verify that emissions of specific constituents of concern are within acceptable limits. A limited set of offgas data is logged on a continuous basis as an indication of the effectiveness of the overall system in destroying or removing organics. Typical offgas parameters that are continuously monitored are levels of carbon dioxide, carbon monoxide, oxygen, and total hydrocarbons.

4.3.5 Quality Assurance/Quality Control

4.3.5.1 Ex-Situ Melters

Quality assurance is applicable to three areas of ex-situ melter operation:

- accuracy and precision of measuring instrumentation;

- sample analyses; and

- procedure and documentation control.

The level of accuracy and precision and the frequency of calibration for each instrument should be defined by the intended use of the data. Critical operating parameters should be measured by instrumentation calibrated according to industry standards and procedures. Accuracy and precision should be within 1 to 2% of the true measurement. Calibration should be performed at least annually. Samples should be analyzed using approved procedures, calibrated instruments, and chemical standards traceable to national standards. Laboratory performance should also be monitored using statistical control charting techniques to monitor short-term and long-term laboratory error. Maintaining a visible procedure and document control process demonstrates that operations, operator training, and roles, responsibilities, authorities, and accountabilities are clearly defined and communicated. Certain sampling and analysis procedures must be traceable to US EPA protocols. For instance, laboratory physical and chemical methods should follow guidelines in US EPA manual SW-846 (40 CFR 60). Offgas aerosol and particulate sampling should follow US EPA Method 5 (US EPA 1986a).

4.3.5.2 In Situ Vitrification

Data necessary to meet QA/QC and data quality objectives often include gas sampling data, scrubber solution analyses, glass sample leaching results, and glass monolith coring/excavation observations. Gas, glass, and scrubber solutions are sampled and analyzed in accordance with established US EPA protocols. Coring in the overlap area between melt settings and excavation at the edge of the remediated site verifies completeness of the melt overlap and that melting depth and width objectives have been met.

5

CASE HISTORIES

This chapter presents an evaluation of case histories deemed pertinent for each of the innovative stabilization/solidification technologies addressed in this monograph.

5.1 Aqueous Stabilization/Solidification

Pilot- and full-scale projects have been completed using several of the innovative aqueous S/S and stabilization processes described. Most S/S applications, however, have relied on conventional equipment designs and operational methods (US EPA 1986b; US EPA 1989c), and so the value of case histories is primarily in the technical results and properties of the treated wastes, which have already been described in this monograph.

There is one area, however, where case histories are especially instructive — auger-type, in situ stabilization and solidification. A small number of pilot- and full-scale projects have been completed using this system (US EPA 1991c; Morse and Dennis 1994). One of special interest is a demonstration project conducted on mixed waste (radioactive and RCRA hazardous) at the DOE Gaseous Diffusion Plant at Portsmouth, Ohio, in 1992 (Benda 1992). This site had soils contaminated with VOCs and low levels of uranium. A full-scale demonstration was performed by Millgard Environmental Corporation for Rust Remedial Services Inc. (then Chemical Waste Management, Inc., ENRAC Group), under the supervision of Martin Marietta Energy Systems.

The project included a number of discrete in situ operations — VOC destruction, VOC removal, and stabilization using the MecTool® auger system described in Section 2.1.1.3. Tests were conducted within a 10 m (33 ft) by 29 m (95 ft) area of the contaminated site. Twelve soil columns were treated in situ to depths of 4.6 m (15 ft) and one column to a depth of 6.7 m (22 ft). The treatment process took from one to four hours per column, depending on the treatment used. Initial VOC concentrations in the soil ranged from 300 to 1,800 mg/L. Each treatment was successful and conferred specific advantages.

The main contamination was trichloroethylene (TCE) in concentrations up to 100 mg/L, with other halocarbons present, along with trace- to low-levels of lead, chromium, uranium 235, and technicium 99 (Siegrist et al. 1993). While there were considerable variations in strength due to non-uniform mixing, solidification was rapidly accomplished and acceptably low TCLP leaching levels of all of the target contaminants were attained.

5.2 Polymer Stabilization/Solidification

5.2.1 Polyethylene Microencapsulation Using Single-Screw Extrusion

Single-screw extrusion for polyethylene microencapsulation of radioactive, hazardous, and mixed waste has successfully progressed from bench-scale process development through full-scale demonstration. A production-scale technology demonstration, sponsored by the of DOE's Office of Technology Development, was conducted at the Brookhaven National Laboratory's Polymer Encapsulation Demonstration and Test Facility (Kalb and Lageraaen 1994, 1996; Kalb et al. 1995b).

The demonstration included all facets of an integrated process necessary to process waste under actual plant conditions. A schematic of the process components is shown in Figure 5.1 and a photo of the facility is shown in Figure 5.2. An aqueous salt waste surrogate containing 35% (by weight) dissolved solids was pretreated to dryness using an indirectly-heated, stirred vacuum dryer and particle size reduction system capable of processing 757 L/day (200 gal/day). The pretreatment system was discussed in Section 3.2.5. Pretreated waste and polymer materials were remotely transferred to storage hoppers by a HEPA-filtered pneumatic transfer system. Production-scale loss-in-weight feeders metered waste to the extruder.

Figure 5.1
Schematic Diagram of the Integrated Polyethylene Encapsulation Process

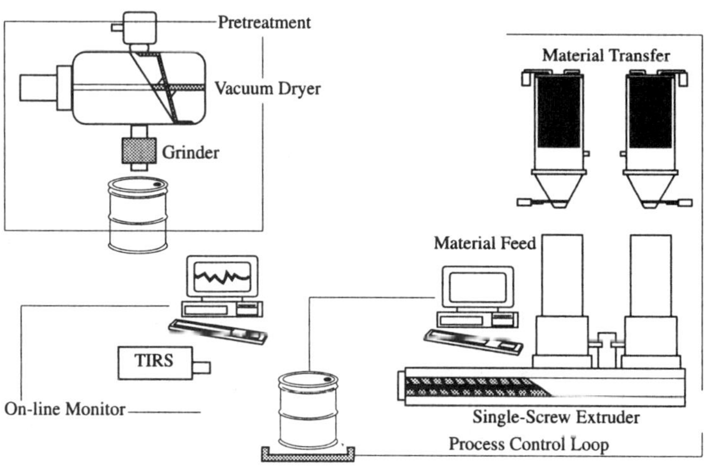

These feeders are capable of extremely precise metering, with accuracies typically around ±0.5%. A 11 4 cm (4.5 in.) single-screw, vented extruder with a maximum output rating of 907 kg/hr (2,000 lb/hr) was used for processing. Extruder output rates were monitored, and data were fed to a computerized process control system that automatically coordinated feeder input rates with extrusion output. A real-time, on-line monitor, developed by Ames Laboratory determined actual waste loadings of the product as it exited the extruder. Process monitoring and instrumentation is described in Section 3.2.7. Typical production-scale processing data for polyethylene microencapsulation using a single-screw extrusion process are provided in Table 5.1.

5.2.2 Polyethylene Macroencapsulation

Envirocare of Utah has been engaged by DOE to commercialize polyethylene macroencapsulation and treat up to 226,800 kg (500,000 lb) of mixed waste lead currently stored throughout the DOE complex. Envirocare is a NRC-licensed disposal facility for naturally-occurring radioactive materials

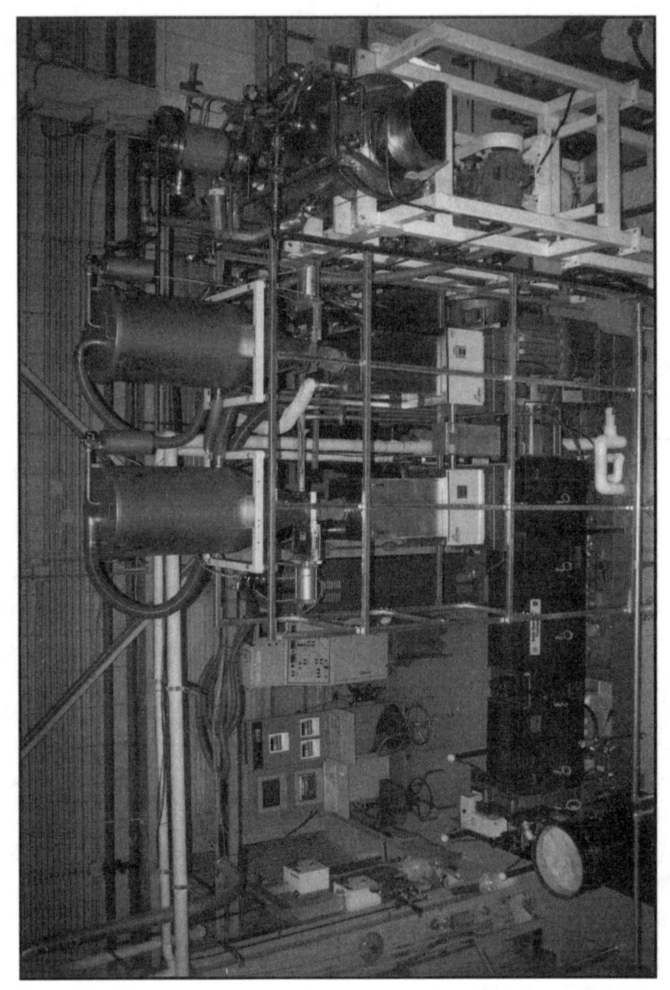

Figure 5.2
Full-Scale Polyethylene Encapsulation Facility

Table 5.1

Production-Scale Process Data for Polyethylene Microencapsulation of Nitrate Salt Wastes Using a Single-Screw Extrusion Process[a]

Parameter		Data	
Heat Zone Temperature Settings:	Zone 1:	280°F	
	Zone 2:	285°F	
	Zone 3:	295°F	
	Zone 4:	325°F	
	Zone 5:	350°F	
Vacuum Pump:		15 mm Hg	
Maximum Screw Speed:		80 rpm	
Maximum Output Rate:		5.4 kg/min	
Waste Loading Range:		30, 40, 50, 60% (by weight)	
Drum Fill Times (55 gallon drum; 60% (by weight) waste loading):	50 rpm	3.5 kg/min	80 min
	80 rpm	5.4 kg/min	50 min

[a]Data from BNL Production-Scale Polyethylene Encapsulation Technology Demonstration

Source: Kalb and Lageraaen 1994

and low-level radioactive wastes. It is the only licensed disposal facility in the U.S. for mixed radioactive/RCRA hazardous wastes. Envirocare conducted a technology demonstration of the polyethylene macroencapsulation process on November 28, 1995, for members of DOE and the local community. Both nonradioactive and radioactive lead brick samples were successfully treated. The lead bricks were packaged in steel cages and suspended in five-gallon metal buckets. A 11.4 cm (4.5 in.) Davis-Standard single-screw plastics extruder was used to extrude clean plastic around the lead to form a layer ≥5.1 cm (2 in.) on all sides. A schematic of the macroencapsulation process is shown in Figure 5.3 along with a schematic of the waste form in Figure 5.4.

Figure 5.3
Single-Screw Extruder for Polyethylene Macroencapsulation

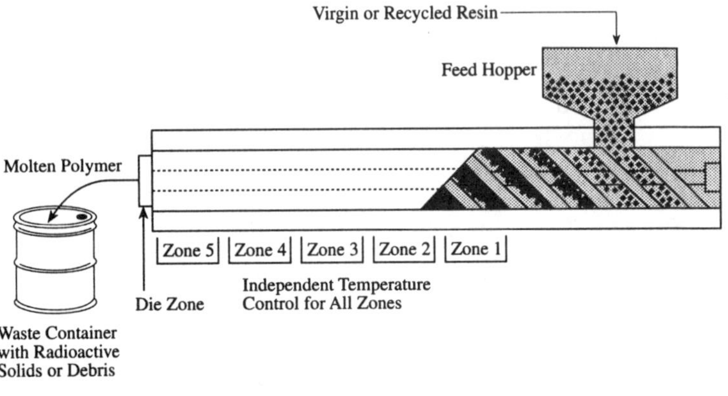

Figure 5.4
Macroencapsulated Waste Form

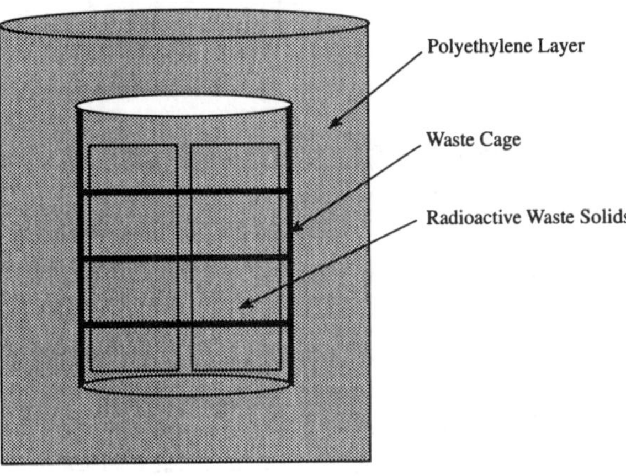

5.2.3 Sulfur Polymer Encapsulation

Idaho National Engineering Laboratory has conducted several pilot-scale tests of sulfur polymer cement (SPC) encapsulation of ash and debris waste (see Section 3.2.3.4). Scientific Ecology Group is using SPC encapsulation to treat incinerator ash residues resulting from their commercially-licensed radioactive waste volume-reduction incinerator facility. Because of the large concentration of contaminants, the ash residues contain significant concentrations of toxic metals in addition to radionuclides.

Scientific Ecology Group modified a high-shear mixing vessel, which is heated by steam for the process. Typically, about 900 kg (2,000 lb) of SPC is heated and melted and approximately 340 kg (750 lb) of incinerator fly ash is added and mixed to form a homogeneous mixture for a waste loading of about 28% (by weight). Currently, Scientific Ecology Group is working with Brookhaven National Laboratory as part of a Cooperative Research and Development Agreement to optimize and demonstrate process applicability to other waste streams.

5.3 Vitrification

5.3.1 Ex-Situ Melters

Babcock & Wilcox (B&W) participated in a US EPA SITE soil treatment demonstration of the B&W cyclone vitrification furnace (Czuczwa et al. 1993). The pilot cyclone furnace was operated between 1990 and 1992 to treat wet and dry contaminated soils (i.e., synthetic soil matrix). The synthetic soil matrix (SSM) was combined with known quantities of heavy metals, organic contaminants, and nonradioactive isotopes of strontium, bismuth, and zirconium. The goals of the testing were as follows.

- determine SSM properties;

- establish cyclone operability with dry soil processing (e.g., feeding, melting behavior, operational data);

- determine slag leachability and volume reduction;

- determine preliminary heavy metals mass balance for the cyclone treatment process;

- design a wet soil feed system and atomizer; and

- establish cyclone operability with wet soil processing.

The pilot cyclone test facility is shown in Figure 5.5. The furnace is a 6 MBtu/hr pilot cyclone furnace, which is water-cooled and simulates the geometry of B&W's front-wall fired, cyclone coal-fired boilers. The furnace is fired by a single, scaled-down version of a commercial coal combustion cyclone furnace. Both the primary and secondary air were heated to approximately 437°C (820°F). For the SITE demonstration, natural gas and preheated primary combustion air entered tangentially into the cyclone burner. In dry soil processing, preheated secondary air, the soil matrix, and a portion of the natural gas entered underneath the secondary air and parallel to the cyclone barrel axis. For wet soil processing, an atomizer sprayed the soil paste directly into the furnace.

Testing was conducted in two phases. In Phase I, dry SSM was processed and in Phase II, wet SSM was processed. Test conditions are shown in Tables 5.2 and 5.3. The B&W system is capable of processing different waste streams under varying conditions of fuel type, water content, and feed rate. The soil matrix was spiked with 7,000 mg/L lead, 1,000 mg/L cadmium, and 1,500 mg/L chromium. Spiked SSM samples were submitted wet and dry for TCLP testing to verify that the starting soil failed the TCLP. The leachability of the lead averaged 81 mg/L; cadmium, 40 mg/L; and chromium, 2.8 mg/L. With the exception of chromium, the spiked solid exceeded US EPA limits for lead (5 mg/L) and cadmium (1 mg/L). Based on feed and product sample composition analyses, test results showed 95 to 97% of the noncondensable portion of the feed was incorporated within the slag product. At processing rates of 23 to 68 kg/hr (50 to 150 lb/hr) dry feed and 45 to 136 kg/hr (100 to 300 lb/hr) wet feed, the heavy metals were retained within the release limits defined under the US EPA TCLP leaching protocol.

During Phase I testing, the cyclone operation remained very stable. Soil input was increased from 21 to 64 kg/hr (46 to 141 lb/hr) with test durations of 3 to 6 hr. The slag tapped well, and no buildup of deposits was observed in the furnace. For Phase II tests, the cyclone was operated at a nominal load of 5 MBtu/hr and 1% excess oxygen. The SSM input varied between 45 and 136 kg/hr (100 and 300 lb/hr). Cyclone operation was characterized

Figure 5.5
B&W Pilot Cyclone Test Facility

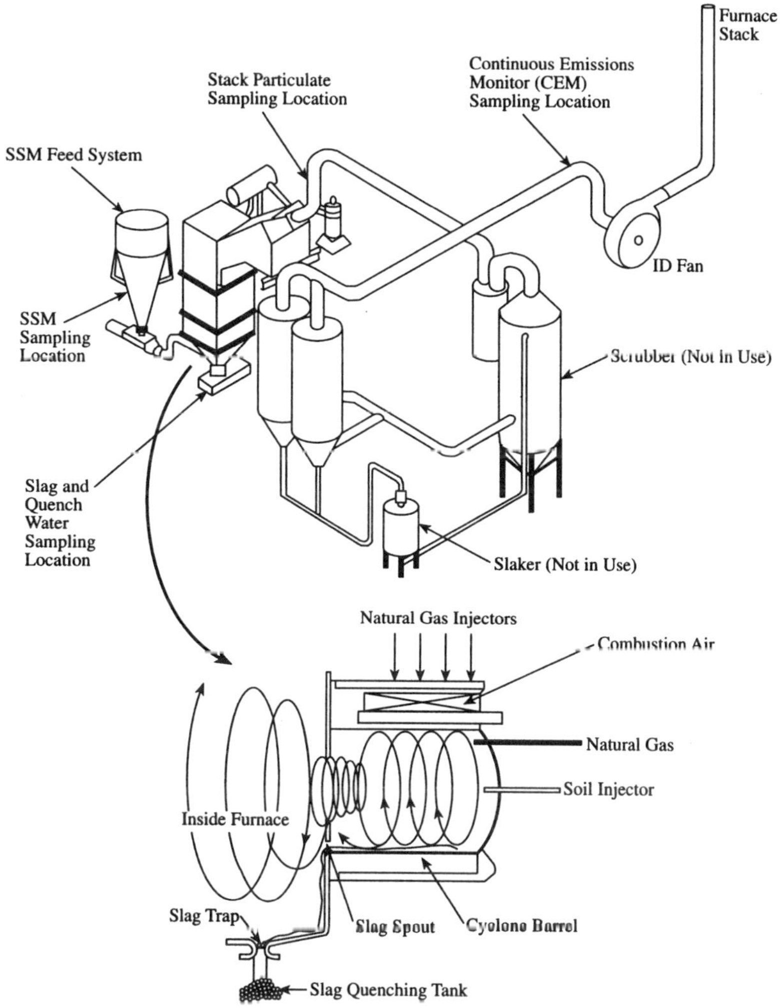

Reproduced courtesy of Babcock Wilcox (1997)

Table 5.2
Phase I Test Matrices

Test	Cyclone Load (MBtu/hr)	SSM Feed Rate (lb/hr)	Stack % Excess Oxygen	Slag Temperature (°F)	Fly Ash/ Slag (%)	Primary and Secondary Air Temperature (°F)
Preliminary Vitrification Tests (Dry, Clean Soil)						
10/25/90	4.8	50	1.1	2340		813
10/26/90	4.6	100	0.8	2430	< 5*	821
10/26/90	4.9	150	0.5	2370		824
10/26/90	4.7	200	0.7	2380		823
Heavy Metals Tests (Dry, Spiked Soil)						
11/01/90	4.6	100	0.7	2350		
11/15/90**	4.7	46	0.5	2400	7.5***	830
11/16/90	4.8	141	0.7	2375	5.9***	817
11/19/90	4.6	94	0.9	2390	5.8***	826

*Amount of the SSM leaving the furnace as ash, preliminary estimate.
**Tests used for TCLP and heavy metals mass balance.
***Includes estimate of amount of particulate deposited in the convection pass.

Source: Czuczwka et al. 1993

as relatively smooth. It was noted during testing that SSM particles smaller than about 1 cm (0.4 in.) could be expected to melt. Larger particles stay in the cyclone until they melt or are carried out by the slag and are encapsulated upon cooling. Unless crushing equipment precludes the presence of large particles, the extent of encapsulated particles which can be tolerated essentially determines the maximum processing rate of the melter unit.

Fluxing agents that cause the soil to melt and flow at lower temperatures could result in decreased metals volatility. To evaluate this possibility, borax (sodium borate decahydrate) was added to the SSM (10% by weight) during one of the Phase II tests. After borax was added, the cyclone load could be reduced to 4.1 MBtu/hr without any problem with slag discharge. With the added borax, the slag temperature was reduced from 1,332°C (2,430°F) at an SSM feed rate of 91 kg/hr (200 lb/hr) to 1,271°C (2,320°F). NO_x levels decreased 20%. Fly ash production increased slightly by 3.5% of the input SSM, presumably due to vaporization of sodium and boron.

Table 5.3
Phase II Test Matrices

Test	Cyclone Load (MBtu/hr)	SSM Feed Rate (lb/hr)	Stack % Excess Oxygen	Slag Temperature (°F)	Fly Ash/ Slag (%)	Primary and Secondary Air Temperature (°F)
		Preliminary Vitrification Tests (Wet, Clean Soil)*				
8/20/91	5.1	100	0.77	2455	1.77	814
8/21/91	5.3	100	0.76	2420	2.01	794
8/27/91	5.0	100	0.62	2370	1.73	814
8/27/91	4.9	150	0.56	2410	0.4	814
8/28/91	4.9	200	0.58	2370	1.91	810
8/28/91	4.9	300	0.53	2390	1.49	813
8/29/91	4.9	300	0.56	2405	1.89	810
9/03/91	5.1	300	0.47		1.94	807
		Heavy Metals Tests (Wet, Spiked Soil)**				
9/09/91	4.8	200	1.2	2430	2.32	823
9/09/91	4.9	200	1.0	2420	2.63	825
9/10/91	4.9	100	0.6	2470		813
9/10/91	4.9	300	0.7	2400		822
9/11/91	4.1***	200	4.7	2320	3.53	810

*Atomizer air 90 to 130 lb/hr, 15 to 100 psig static pressure.
**Atomizer air flow rates of 128 to 134 lb/hr were used.
***10% Borax was added to the SSM.

Source: Czuczwka et al. 1993

Testing results indicated that the process would be well-suited for the treatment of low volatility contaminants, such as many radionuclides. At least 95 to 97% of the input SSM was incorporated into the slag. The heavy metals partitioned between the vitrified slag and the stack fly ash. The capture of heavy metals in the vitrified slag from all tests ranged from 8 to 17% (by weight) for cadmium, 24 to 35% (by weight) for lead and 80 to 95% (by weight) for chromium. The capture of heavy metals in the slag increased with increasing feed rate and with decreasing metal volatility. Stable cyclone operation was achieved during both phases of testing. Concentrations of CO and NO_x in the offgas were within acceptable ranges. Soil volumes were reduced between 25 and 35% (dry basis) through vitrification.

As a result of the SITE demonstration, several recommendations for technology application were made. It is believed that the combustion melting technology is best suited for treating soils contaminated by organics and either very high or very low volatility heavy metals. The high heat release rates and turbulence make the cyclone vitrification process well-suited for organics destruction. Vitrification of very high-volatility metals or radionuclides could tend to concentrate those elements in the relatively small fly ash stream, from which they could be recovered. Vitrification of very low-volatility metals or radionuclides would tend to concentrate those elements in a nonleachable waste form. For intermediate volatility metals or radionuclides, recycling the fly ash to the melter is expected to be the best option.

5.3.2 In Situ Vitrification

The findings associated with a demonstration of the Geosafe Corporation (Geosafe 1995) in situ vitrification (ISV) process under the US EPA SITE Program in conjunction with remediation activities associated with an US EPA Region V removal action. The technology was assessed for its ability to reduce pesticides (specifically chlordane, dieldrin, and 4,4'-DDT) and mercury to below Region V-mandated limits. It was evaluated against the nine criteria for decision-making in the Superfund Feasibility Study process. Table 5.4 presents the results of this evaluation.

As part of the Region V removal action, Geosafe performed a total of eight melts that covered nine pre-staged treatment cells at the Parsons Chemical Works, Inc. site located in Grand Ledge, Michigan. The SITE

Program studied one of these treatment settings (Cell 8) in detail to determine the technology's ability to meet the Region V removal criteria and to obtain cost and performance data on the technology.

Results for the treated soil are based on posttreatment sampling just below the surface of the melt alone. Complete posttreatment sampling of the solidified melt could not be safely performed until at least one year after treatment, at which time sampling of the melt core will take place. Because the technology is already being used in commercial applications, a report (Geosafe 1995) has been published prior to obtaining treated soil samples from the center of the study area. In this manner, the community is provided with the information currently available regarding the operability and effectiveness of the technology. Results of the posttreatment soil samples collected from the core of Cell 8 will be reported at a later date in a published addendum.

Conclusions Based on Critical Objectives. The studies conducted by the SITE Program suggest the following conclusions regarding the technology's performance at the Parsons' site based on the critical objectives stated for the demonstration.

- The treated soil met the US EPA Region V cleanup criteria for the target pesticides and mercury. Dieldrin and 4,4'-DDT were reduced to levels below their analytical reporting detection limits (<16 µg/kg) in the treated soil. Chlordane was below its detection limit (80 µg/kg) before treatment commenced. Mercury, analyzed by standard SW-846 Method 7471 procedures, was below the specified cleanup level before treatment began, averaging 3,800 µg/kg. It was reduced by volatilization to an average of less than 33 µg/kg in the treated soil.

- Stack gas samples were collected during the demonstration to characterize process emissions. No target pesticides were detected in the stack gas samples. During the demonstration, mercury emissions averaged $5.4 \cdot 10^4$ µg/hr ($1.2 \cdot 10^{-4}$ lb/hr). The emissions were below the regulatory requirement of $2.7 \cdot 10^5$ µg/hr ($5.93 \cdot 10^{-4}$ lb/hr) at all times. Other metal emissions in the stack gas (particularly arsenic, chromium, and lead) were of regulatory concern during process operations, but were found to be in compliance with Michigan's applicable or relevant and appropriate requirements (ARARs).

Table 5.4
Evaluation Criteria for the Geosafe In Situ Vitrification Process

Overall Protection of Human Health and the Environment	Compliance with ARARs	Long-Term Effectiveness and Performance	Short-Term Effectiveness
Provides both short- and long-term protection by destroying organic material. Developer also claims the technology can treat radioactive compounds. Remediation can be performed in situ, thereby reducing the need for excavation. Offgas treatment system reduces airborne emissions. System is flexible and can be adapted for a variety of contaminant types and site conditions. Technology can simultaneously treat a mixture of waste types. Technology is applicable to combustible materials, but the concentration of such materials in the treatment zone must be carefully controlled and treatment prudently planned.	Requires compliance with RCRA treatment, storage, and land disposal regulations (for a hazardous waste). Successfully treated waste may be delisted or handled as nonhazardous waste. Operation of on-site treatment unit may require compliance with location-specific applicable or relevant appropriate requirements (ARARs). Emission control may be needed to ensure compliance with air quality standards depending upon local ARARs and test soil components. Scrubber water will likely require secondary treatment before discharging to publicity owned treatment works (POTW) or surface bodies. Disposal requires compliance with Clean Water Act regulations.	Effectively destroys organic contamination and immobilizes inorganic material. Developer also claims the technology can treat radioactive compounds. Reduces the likelihood of contaminants leaching from treated soil. ISV glass is thought to have a stability similar to volcanic obsidian which is estimated to remain physically and chemically stable for thousands to millions of years. Allows potential reuse of property after treatment.	Treatment of a site using ISV destroys organic compounds and immobilizes inorganic contaminants. Vitrification of a single treatment setting may be completed in approximately ten days. This treatment time may vary depending on-site-specific conditions. Presents potential short-term chemical exposure risks to workers operating process equipment. High voltage and high temperatures require appropriate safety precautions. Some short-term risks associated with air emissions are dependent upon test material composition and offgas treatment system design. Staging, if required, involves excavation and construction of treatment areas. A potential for fugitive emissions and exposure exists during excavation and construction.

Reduction of Toxicity, Mobility, or Volume through Treatment	Implementability	Cost	Community Acceptance	State Acceptance
Significantly reduces toxicity and mobility of soil contaminants through treatment. Volume reductions of 20 to 50% are typical after treatment. Some inorganic contaminants, especially volatile metals, may be removed by the vitrification process, and require subsequent treatment by the offgas treatment system. Some treatment residues may themselves be treated during the next vitrification setting. Residues from the final setting, including expended or contaminated processing equipment, may require special disposal requirements. Volume of scrubber water generated is highly dependent upon soil moisture content.	Equipment is mobile and can be brought to a site using conventional shipping methods. Weight restrictions on tractors/trailers may vary from state to state. Support equipment includes earth moving equipment for staging treatment areas (if required) and covering treated areas with clean soil. A crane is required for offgas containment hood placement and movement. Chemical characterization of contaminated soil is required for proper offgas treatment system design. A suitable source of electric power is required to utilize this technology. Technology not recommended for sites which contain organic content greater than 7 to 10% (by weight), metals content in excess of 25% (by weight), and inorganic contaminants greater than 20% (by volume). Sites with buried drums may only be treated if drums are not intact or sealed.	The cost for treatment when the soils staged into 15 ft deep cells is approximately $770/yd^3 ($430/ton). Treatment is most economical when treating large sites to the maximum depth. Electrical power is generally the most significant cost associated with ISV. Other factors (in order of significance) include labor costs, startup and fixed costs, equipment costs, and facilities modifications and maintenance costs. Moisture content of the medium being treated directly influences the cost of treatment since electric energy must be used to vaporize water before soil melting occurs. Sites that require staging and extensive site preparation will have higher overall costs.	Technology is generally accepted by the public because it provides a permanent solution and because it is performed in situ. Potential reuse of land after treatment provides an attractive alternative to property owners. A public nuisance could be created if odorous emissions from the soil constituents are not properly controlled by the offgas system.	State ARARs may be more stringent than federal regulations. State acceptance of the technology varies depending upon ARARs. The ISV system (especially the offgas treatment portion) is somewhat modular, such that it may be modified to meet state-specific criteria.

- Emissions of total hydrocarbons (as propane) and carbon monoxide are regulated at 100 ppm_v and 50 ppm_v, respectively. Throughout the demonstration, vapor emissions of these gases (measured downstream from the thermal oxidizer) were well below the regulatory guidelines. Total hydrocarbon and CO emissions both averaged below 10 ppm_v.

Conclusions Based on Secondary Objectives. The studies conducted by the SITE Program suggest the following conclusions regarding the technology's performance at the Parsons' site based on the secondary objectives stated for the demonstration.

- The technology successfully treated the soil in Cell 8, completing the test cell melt in ten days with only minor operational problems. During this time, approximately 252 m^3 (330 yd^3) (approximately 540 tonne [600 ton]) of contaminated soil were vitrified, according to Geosafe melt summaries. Approximately 610 MWhr (2,080 MBtu) of energy was applied to the *total* soil volume melted (estimated to be 367 m^3 [480 yd^3]) during vitrification of Cell 8; power applied to the actual *contaminated* soil volume could not be independently measured because clean fill and surrounding uncontaminated soil were vitrified as part of each melt. Based on the total soil treated in Cell 8, the energy consumption was approximately 0.72 MWhr/ton (2.5 Btu/ton). System operation was occasionally interrupted briefly for routine maintenance such as electrode segment addition and adjustment.

- The solid, vitrified material collected was subjected to TCLP analysis for the target pesticides and mercury. Test results indicated that no target pesticides were detected in the posttreatment leachate. Chlordane was not detected in either the pre- or post-treatment leachate, so no definitive conclusions can be drawn about the technology's impact on the leachability of this compound based on this demonstration. Levels of leachable mercury in both pre- and posttreatment soil leachates were well below the regulatory limit of 200 μg/L (40 CFR §261.24). Several other metals were also found to have passed the TCLP leaching test.

- Scrubber water generated during the demonstration contained volatile organics, partially-oxidized semivolatile organics (phenolics), mercury, and other metals. The scrubber water underwent secondary treatment before ultimate disposal, and data suggest that secondary treatment of this waste stream will probably be required in most cases.

- Pretreatment soil *dry* density averaged 1.8 tonne/m³ (1.5 ton/yd³), while posttreatment soil *dry* density averaged 2.5 tonne/m³ (2.1 ton/yd³). Accordingly, a volume reduction of approximately 30% was observed for the test soil on a dry basis.

- The cost for treatment when the soil is staged into nine cells is approximately $815/tonne ($740/ton) or $1,300/yd³ for 1.5 m (5 ft) deep cells, $474/tonne ($430/ton) or $770/yd³ for 4.6 m (15 ft) deep cells (like those at the Parsons' site), and $407/tonne ($370/ton) or $660/yd³ for 6.1 m (20 ft) deep cells. The costs presented are calculated based on the number of cubic yards of *contaminated* soil treated. Because clean fill and surrounding uncontaminated soil are treated as part of each melt, the *total* amount of material treated is greater than the amount of contaminated soil treated. Costs per cubic yard based on total soil treated would, therefore, be lower than the costs per cubic yard based on contaminated soil treated presented in this report.

- Treatment is most economical when treating large cells to the maximum depth. The primary cost categories include utilities, labor, start-up, and fixed costs.

The site studied during this demonstration was Geosafe's first large-scale commercial project. Valuable lessons learned at this site have been put into practice in subsequent applications.

LIST OF REFERENCES

Adams, J.W. and P.D. Kalb. 1996. Thermoplastic stabilization of a chloride, sulfate, and nitrate salts mixed waste surrogate. *Emerging Technologies in Hazardous Waste Management VII.* In Press.

American National Standards Institute/American Nuclear Society (ANSI/ANS). 1986. American National Standard for measurement of the leachability of solidified low-level radioactive wastes by short-term test procedure. *ANSI/ANS.* 16(1). April 14.

American Society of Testing Materials (ASTM). 1970. Standard recommended practice for determining resistance of synthetic polymeric materials to fungi. ASTM G-21. New York, NY. April.

American Society of Testing Materials (ASTM). 1970. Tentative recommended practice for determining resistance of plastics to bacteria. ASTM G-22. New York, NY.

American Society of Testing Materials (ASTM). 1990. *Standard Test Method for Flow Rates of Thermoplastics by Extrusion Plastomer.* ASTM D1238-90b. New York, NY. December.

Barth, E.F., M. Taylor, J. Wenz, and S. Giti-Pour. 1995. Extraction, recovery, and immobilization of chromium from contaminated soils. *Proceedings of the 49th Industrial Waste Conference.* Boca Raton, FL: Lewis Publishers.

Benda, G. 1992. *Commercial Experience in Treating U.S. Department of Energy Mixed Wastes.* Columbia, SC: Rust Remedial Services, Inc.

Bostick, W.D., D.P. Hoffmann, J.M. Chiang, W.H. Hermes, L.V. Gibson, Jr., A.A. Richmond, J. Mayberry, and G. Frazier. 1993. *Surrogate Formulations for Thermal Treatment of Low-Level Mixed Waste: Part II: Selected Mixed Waste Treatment Project Waste Streams.* DOE MWIP-16. Oak Ridge, TN: Martin Marietta Energy System. September.

Bounini, L. 1990. *Stabilization/Solidification: Advances in Process Engineering.* Pacific Basin Consortium for Hazardous Waste Research, 1990 Honolulu Conference. Honolulu, Hawai. November.

Buelt, J.L., C.L. Timmerman, K.H. Oma, V.F. Fitzpatrick, and J.G. Carter. 1987. *In Situ Vitrification of Transuranic Waste: An Updated Systems Evaluation and Applications Assessment.* PNL-4800 Suppl. 1. Richland, WA: Pacific Northwest Laboratory.

Callow, R.A., L.E. Thompson Weidner, C.A. Loehr, B.P. McGrail, and S.O. Bates. 1991. *In Situ Vitrification Application to Buried Waste: Final Report of Intermediate Field Tests at Idaho National Engineering Laboratory.* EGG-WTD-9807. Richland, WA: Pacific Northwest Laboratory.

Calmus, R.B. and L.R. Eisenstatt. 1995. *High-Level Melter Alternatives Assessment - Compilation of Information Generated During Assessment.* WHC-MR-0490. Richland, WA: Westinghouse Hanford Company.

Cambell, B.E., S. Schultz, and J. Cichelli. 1994. Testing of in situ vitrification on soils contaminated with explosive compounds. *Proceedings of Federal Environmental Restoration Three — Waste Minimization Two.* New Orleans, LA. April 22-29.

Carter, J.G., S.S. Koegler, and S.O. Bates. 1988. *Process Performance of the Pilot-Scale In Situ Vitrification of a Simulated Waste Disposal Site at the Oak Ridge National Laboratory.* PNL-6530. Richland, WA: Pacific Northwest Laboratory.

Chapman, C.C. 1991. Evaluation of vitrifying municipal incinerator ash. *Ceramic Transactions, Nuclear Waste Management IV.* 23: 223-232. Cincinnati, OH: American Ceramic Society.

Chapman, C.C. and J.L. McElroy. 1989. Slurry-fed ceramic melter - a broadly accepted system to vitrify high-level waste. *High Level Radioactive Waste and Spent Fuel Management Vol. II.* New York, NY: American Society of Mechanical Engineers. pp 119-127.

Colombo, P. and R.M. Neilson. 1979. *Properties of Radioactive Wastes and Waste Containers, First Topical Report.* BNL-NUREG-50957. Upton, NY: Brookhaven National Laboratory. August.

Colombo, Peter, Edwin Barth, Paul L. Bishop, Jim Buelt, and Jesse R. Conner. 1994. *Innovative Site Remediation Technology — Stabilization/Solidification.* Annapolis, MD: American Academy of Environmental Engineers.

Conner, J.R. 1990. *Chemical Fixation and Solidification of Hazardous Wastes.* New York: Van Nostrand Reinhold.

Conner, J.R. 1991. Treatment of D007 inorganic solid debris containing soluble chrome. 93rd Annual Meeting of the American Ceramics Society. Cincinnati, OH. April 29.

Conner, J.R. 1992. The role of stabilization/solidification in hazardous waste disposal. Proc. HazMat '92 International. Atlantic City, NJ.

Conner, J.R. 1997. *Guide to Improving the Effectiveness of Cement-Based Stabilization/Solidification.* Portland Cement Association, Skokie, IL.

Conner, J.R. and P.R. Lear. 1991. *Immobilization of low-level organic compounds in hazardous wastes.* Air & Waste Management Association, 84th Annual Meeting & Exhibition. Vancouver, British Columbia, Canada. June 16-21.

Conner, J.R. and R.S. Reber. 1992. Australian Patent 615459. January 31.

Conner, J.R. and F.G. Smith. 1993. Immobilization of low-level hazardous organics using recycled materials. Third International Symposium on Stabilization/Solidification of Hazardous, Radioactive, and Mixed Wastes. Williamsburg, VA. November 1-5.

Czuczwa, J.M., J.J. Warchol, W.F. Musiol, and H. Farzan. 1993. *SITE Emerging Technologies Project: Babcock & Wilcox Cyclone Vitrification.* EPA/540/R-93/507. Cincinnati, OH: US EPA.

Darnell, G.R. 1991. Sulfur polymer cement, a new stabilization agent for mixed and low- level radioactive waste. *Proceedings of the First International Symposium on Mixed Waste.* A.A. Moghissi and G.A. Benda (eds.). Baltimore, MD: University of Maryland.

Darnell, G.R. 1993. Non-radioactive full-scale tests with sulfur polymer cement for stabilization of incinerator ash. Fifteenth Annual U.S. Department of Energy Low-Level Radioactive Waste Management Conference. December 1.

Davis-Standard Corporation. *Knowing your Extruder.* Manufacturers Product Literature. Pawcatuck, CT: Crompton and Knowles Corp.

Diel, B.N., D.J. Kuchynka, and J. Borchert. 1995. Complexed metals in hazardous waste: limitations of conventional chemical oxidation. Superfund XV Conference.

Dow Chemical Company. 1978. The Dow system for solidification of low-level radioactive waste from nuclear power plants. Topical Report. Midland, MI: Dow Chemical Co. March.

Draiswerke, Inc. *Drais News.* (1)4. Manufacturers Product Literature. Mahwah, NJ.

Durham, R.L. and C.R Henderson. 1984. U.S. Patent 4,460,292. July 17.

Eighmy, T.T., S.F. Bobouski, T.B. Ballestero, and M.R. Collins. 1991. *Investigation into Leachate Characteristics from Amended and Unamended Combined Ash and Scrubber Residues.* Unpublished report to Concord N.H. Regional Solid Waste/Resource Recovery Cooperative. University of New Hampshire, Durham, NH.

Environmental Technologies Alternatives, Inc. 1994a. Immobilization of organics in debris and characteristic (D-code) wastes. Technical Bulletin TB-234. Port Clinton, OH.

Environmental Technologies Alternatives, Inc. 1994b. Control of air emissions in hazardous waste stabilization operations. Technical Bulletin TB-214. Port Clinton, OH.

Environmental Technologies Alternatives, Inc. 1995. KAX test report summary. Technical Bulletin. Port Clinton, OH.

Eyler, L.L., M.L. Elliott, D.L. Lessor, and P.S. Lowery. 1991. Computer modeling of ceramic melters to assess impacts of process and design variables on performance. *Ceramic Transactions, Nuclear Waste Management IV.* 23: 395-408. Cincinnati, OH: American Ceramic Society.

Faucette, A.M., B.A. Logsdon, J.J. Lucerna, and R.J. Yudnich. 1994. Polymer solidification of mixed wastes at the rocky flats plant. *Waste Management '94, Volume 3.* R.G. Post (ed.). Proceedings of the Symposium on Waste Management. Tucson, AZ. February 27-March 3.

Faucette, A.M., R.H. Getty, L.G. Peppers, M.J. Serra, and L.J. Wood. 1995. Thermoset macroencapsulation of surrogate beryllium fines wastes. Waste Mangagement '95 Symposium. Tucson, AZ. February.

Frados, J. (ed.). 1985. Machinery and equipment. *Plastics Compounding Redbook.* 8(6): 108-124.

Franz, E.M., J.H. Heiser, and P. Colombo. 1987. *Solidification of Problem Wastes Annual Progress Report.* BNL-52078. Upton, NY: Brookhaven National Laboratory. February.

Freeman, C.J., S.K. Sundaram, and D.A. Lamar. 1995. Corrosion of electrode materials for a high-level waste, high-temperature melter. *Corrosion of Materials by Molten Glass.* American Ceramic Society and Pacific Northwest National Laboratory, Richland, WA.

Fuhrmann, M. and P.D. Kalb. 1993. Leaching behavior of polyethylene encapsulated nitrate waste. Third International Symposium on Stabilization/Solidification of Hazardous, Radioactive and Mixed Wastes, ASTM. Williamsburg, VA. November 1-5.

Geosafe Corporation. 1995. In situ vitrification innovative technology evaluation report. EPA/540/R-94/520. Cincinnati, OH.

Gilliam, T.M., L.R. Dole, and E.W. McDaniel. 1986. Waste immobilization in cement-based grouts. ASTM Special Technical Publication. Philadelphia, PA: ASTM.

Heiser, J.H. 1995. *In-Situ Stabilization of TRU/Mixed Waste.* TTP CH3-6-LF-23. DOE Office of Technology Development, Washington, DC. October.

Heiser, J.H. and L.W. Milian. 1994. *Laboratory Evaluation of Performance and Durability of Polymer Grouts for Subsurface Hydraulic/Diffusion Barriers.* BN 61292. Upton, NY: Brookhaven National Laboratory. May.

Heiser, J.H., P. Colombo, and J. Clinton. 1992. *Polymers for Subterranean Containment Barriers for Underground Storage Tanks (USTs).* BNL-48750. Upton, NY: Brookhaven National Laboratory. December.

Hnat, J.G., W.F. Olix, W.F. Talley, and L.M. Bartone. 1991. A cyclone melting system for processing hazardous waste dusts. Presented at the National Research and Development Conference on the Control of Hazardous Materials. February 20-22.

Jowett, M. and H.I. Price. 1932. Solubilities of the phosphates of lead. *Trans. Faraday Soc.* Volume 28.

Kalb, P.D. 1993. *Polyethylene Encapsulation of Simulated Blowdown Waste for Seg Treatabiltiy Study.* BNL-49378. Upton, NY: Brookhaven National Laboratory. August.

Kalb, P.D. 1995a. Telephone conversation with D. Sangster, Lex Technologies, Inc. May 25.

Kalb, P.D. 1995b. Sulfur polymer encapsulation of radioactive, hazardous and mixed wastes. US EPA Workshop. November.

Kalb, P.D. and P. Colombo. 1985a. *Modified Sulfur Cement Solidification of Low-Level Wastes, Topical Report.* BNL-51923. Upton, NY: Brookhaven National Laboratory. October.

Kalb, P.D. and P. Colombo. 1985b. *An Economic Analysis of a Volume Reduction/Polyethylene Solidification System for Low-Level Radioactive Wastes.* BNL-518666. Upton, NY: Brookhaven National Laboratory. January.

Kalb, P.D. and P. Colombo. 1997. *Composition and Process for the Encapsulation and Stabilization of Radioactive, Hazardous, and Mixed Wastes.* U.S. Patent 5,649,323. July 15.

Kalb, P.D. and P.R. Lageraaen. 1994. *Polyethylene Encapsulation Full-Scale Technology Demonstration, Final Report.* BNL-52478. Upton, NY: Brookhaven National Laboratory. October.

Kalb, P.D. and P.R. Lageraaen. 1996. Full-scale technology demonstration of a polyethylene encapsulation process for radioactive, hazardous, and mixed wastes. *Journal of Environmental Science and Health.* In press.

Kalb, P.D., J.H. Heiser, and P. Colombo. 1991a. *Polyethylene Encapsulation of Nitrate Salt Wastes: Waste Form Stability, Process Scale-Up, and Economics.* BNL-52293. Upton, NY: Brookhaven National Laboratory. July.

Kalb, P.D., J.H. Heiser, and P. Colombo. 1991b. Modified sulfur cement encapsulation of mixed waste contaminated incinerator fly ash. *Waste Management Journal.* 11: 147-153.

Kalb, P.D., J.H. Heiser, and P. Colombo. 1993. Long-term durability of polyethylene for encapsulation of low-level radioactive, hazardous, and mixed wastes (Chapter 22). *Emerging Technologies in Hazardous Waste Management.* D.W. Tedder and F.G. Pohland (eds.). ACS Symposium Series No. 518. American Chemical Society. April.

Kalb, P.D., J.H. Heiser, R. Pietrzak, and P. Colombo. 1991. Durability of incinerator ash waste encapsulated in modified sulfur cement. Proceedings of the 1991 Incineration Conference. Knoxville, TN. May 13-17.

Kalb, P.D., M.G. Cowgill, L.W. Milian, and E.C. Selcow. 1995a. *BNL Building 650 Lead Decontamination and Treatment Feasibility Study.* Upton, NY: Brookhaven National Laboratory. In press.

Kalb, P.D., P.R. Lageraaen, S.L. Wright, and M. Hill. 1995b. Full-scale integrated technology demonstration of the polyethylene encapsulation process for improved final waste forms. Waste Mangagement '95 Symposium. Tucson, AZ. February.

Koegler, S.S., R.K. Nakaoka, R.K. Farnsworth, and S.O. Bates. 1989. *Vitrification Technologies for Weldon Springs Raffinate Sludges and Contaminated Soils Phase 2 Report: Screening of Alternatives.* PNL-7125. Richland, Washington: Pacific Northwest Laboratory.

Krueger, R.C., A.K. Chowdhury, and M.A. Warner. 1991. Full-scale remediation of a grey iron foundry waste surface impoundment. *Environ. Prog.* 10(3). August.

Lageraaen, P.L., B.R. Patel, and P.D. Kalb. 1995. *Preliminary Treatability Studies for INEL Mixed Wastes.* BNL-62620. Upton, NY: Brookhaven National Laboratory.

Lageraaen, P.R., P.D. Kalb, D.L. Grimmett, R.L. Gay, and C. D. Newman. 1995. *Polyethylene Encapsulation of Molten Salt Oxidation Mixed Low-Level Radioactive Salt Residues.* Proceedings of the Third Biennial ASME Mixed Waste Symposium. Baltimore, MD. August.

Lamar, D.A., M.F. Cooper, and C.J. Freeman. 1995. Demonstration of the high-temperature melter for treatment of hanford tank waste. *Proceedings of the International Symposium on the Environmental Issues and Waste Management Technologies in Ceramic and Nuclear Industry.* American Ceramic Society and Pacific Northwest National Laboratory, Richland, Washington. In press.

Lear, P.R. and J.R. Conner. 1991. Immobilization of low-level organic compounds in contaminated soil. Sixth Annual Conference on Hydrocarbon Contaminated Soils. Amherst, MA. September 23-26.

Lear, P.R. and J.R. Conner. 1992. Treatment of arsenic trisulfide waste by chemical fixation. Presented at 85th Annual Meeting of the Air & Waste Management Association. Kansas City, KS. June 21-26.

Lighty, JoAnn, Martha Choroszy-Marshall, Michael Cosmos, Vic Cundy, and Paul De Percin. 1993. *Innovative Site Remediation Technology — Thermal Desorption.* Annapolis, MD: American Academy of Environmental Engineers.

Lomenick, T.F. 1992. *Proceedings of the Workshop on Radioactive, Hazardous, and/or Mixed Waste Sludge Management.* CONF-901264. Oak Ridge, TN: Martin Marietta Energy Systems. January.

Loomis, G., D. Thompson, and J.H. Heiser. 1995. *Innovative Sub-Surface Stabilization of Transuranic Pits and Trenches.* INEL-95-0632. Idaho Falls, ID: Idaho National Engineering Laboratory. December.

Luey, J., S.S. Koegler, W.L. Kuhn, P.S. Lowery, and R.G. Wintleman. 1992. *In Situ Vitrification of a Mixed-Waste Contaminated Soil Site: The 116-B-6A Crib at Hanford.* PNL-8281. Richland, WA: Pacific Northwest Laboratory.

Magee, Richard S., James Cudahy, Clyde R. Dempsey, John R. Ehrenfield, Francis W. Holm, Dennis Miller, and Michael Modell. 1994. *Innovative Site Remediation Technology — Thermal Destruction.* Annapolis, MD: American Academy of Environmental Engineers.

Mattus, C.H. and A.J. Mattus. 1994. *Evaluation of Sulfur Polymer Cement as a Waste Form for the Immobilization of Low-Level Radioactive or Mixed Waste.* ORNL/TM-12657. March.

Mayberry, J.L., L.M. DeWitt, R. Darnell, R. van Konyenburg, W. Greenhalgh, D. Singh, R. Schumacher, P. Erickson, J. Davis, and R. Nakaoka. 1993. *Technical Area Status Report for Low-Level Mixed Waste Final Waste Forms.* DOE/MWIP-3. August.

McLaughlin, T.L., S.P. Airhart, K.L. Beattie, J.K. Rouse, S.J. Phillips, and W.E. Stewart. 1992. *Interim Subsurface Barrier Technologies Workshop Report.* Prepared for Westinghouse Hanford Company, Richland, WA. September.

Millgard, D.V. and R.H. Kappler. 1992. U.S. Patent 5,135,058. August 28.

Millgard Environmental Corp. 1992. MecTool®. Livonia, MI.

Morse, J.G and D. Dennis. 1994. Assessment and cleanup of soils using in situ stabilization at a manufactured gas plant site. Presented at 87th Annual Meeting & Exhibition, Air & Waste Managment Association. Cincinnati,OH. June 19-24.

Neilson, R.M. and P. Colombo. 1982. *Waste Form Development Program Annual Progress Report.* BNL-51517. Upton, NY: Brookhaven National Laboratory. January.

Neilson, R.M., P.D. Kalb, M. Fuhrmann, and P. Colombo. 1983. Solidification of ion exchange resin wastes in hydraulic cement. *The Treatment and Handling of Radioactive Wastes.* A.G. Blasewitz, J.M. Davis, and M.R. Smith (eds.). New York: Springer-Verlag.

O'Hara, M.J. and M.R. Surgi. 1988. *Immobilization of Lead and Cadmium in Solid Residues from the Combustion of Refuse Using Lime and Phosphate.* U.S. Patent 4,737,536. April 12.

Orfeuil, M. 1987. *Electric Process Heating - Technologies/Equipment/Applications.* Battelle Memorial Institute, Columbus, OH: Battelle Press.

Patel, B.R., P.R. Lageraaen, and P. Kalb. 1995. *Review of Potential Processing Techniques for the Encapsulation of Waste in Thermoplastic Polymers.* BNL-62200. Upton, NY: Brookhaven National Laboratory. August.

Place, B.G. 1990. *Engineering Study for the Treatment of Spent Ion Exchange Resin Resulting from Nuclear Process Applications.* WHC-EP-0375. Richland, WA: Westinghouse Hanford Co. September.

Rysman de Lockerente, S. 1979. *Treatment of Wastes, Especially Toxic Wastes, with a Silicate to Form a Solid Aggregate.* Belgian Patent 842,206.

Shade, J.W., R.K. Farnsworth, J.S. Tixier, and B.L. Charboueau. 1991. *Engineering-Scale Test 4: In Situ Vitrification of Toxic Metals and Volatile Organics Buried in INEL Soils.* PNL-7611. Richland, WA: Pacific Northwest Laboratory.

Shearer, T., J.G. Hnat, J.S. Patten, J.M. Santioanni, and J. St. Clair. 1992. Vitrification of heavy metal contaminated soils with Vortec Corporation's Combustion & Melting System (CMS). Presented at the I&EC Special Symposium, American Chemical Society. Atlanta, GA. September 21-23.

Siegrist, R.L. et al. 1992. In-situ treatment of contaminated clay soils by physicochemical processes coupled with soil mixing: highlights of the x-231B technology demonstration. Presented at I&EC Special Symposium, American Chemical Society. Atlanta, GA. September 21-23.

Siegrist, R.L., S.R. Cline, T.M. Gilliam, and J.R. Conner. 1993. In-situ stabilization of mixed waste contaminated soil. Draft of paper to be presented at an ASTM symposium.

Siskind, B. and J.H. Heiser. 1993. *Regulatory Issues and Assumptions Associated with Barriers in the Vadose Zone Surrounding Buried Waste.* BNL-48749. Upton, NY: Brookhaven National Laboratory. February.

Spalding, B.P., G. K. Jacobs, N.W. Dunbar, M.T. Naney, J.S. Tixier, and T. D. Powell. 1992. *Tracer-Level Radioactive Pilot-Scale Test of In Situ Vitrification for the Stabilization of Contaminated Soil Sites at ORNL.* ORNL/TM-12201. Oak Ridge, TN: Oak Ridge National Laboratory.

Spence, R.D., T.M. Gilliam, I.L. Morgan, and S.C. Osborne. 1990. Stabilization/solidification of wastes containing volatile organic compounds in commercial cementitious waste forms. Presented at 2nd Inter. Symp. on S/S of Haz., Radioactive and Mixed Wastes. Williamsburg, VA. May 29-June 1.

Sundaram, S.K., C.J. Freeman, and D.A. Lamar. 1995. Electrochemical corrosion and protection of electrodes for a high-level, temperature-temperature glass melter. *Corrosion of Materials by Molten Glass.* American Ceramic Society and Pacific Northwest National Laboratory, Richland, WA.

Thompson, L.E., S.O. Bates, and J.E. Hansen. 1992. *Technology Status Report: In Situ Vitrification Applied to Buried Wastes.* PNL-8219. Richland, WA: Pacific Northwest Laboratory.

Thompson, L.E., J.L. McElroy, and C.L. Timmerman. 1995. International applications status of in situ vitrification on mixed TRU and LLW buried wastes. *Proc. of the Third Biennial Mixed Waste Symposium.* Baltimore, MD.

U.S. Department of Energy (DOE). 1988. DOE Order 5820.2A. Radioactive Waste Management. Washington, DC: DOE. September.

U.S. Nuclear Regulatory Commission (NRC). 1983. *Licensing Requirements for Land Disposal of Radioactive Waste.* 10 CFR, Part 61. Washington DC: US NRC. May.

U.S. Nuclear Regulatory Commission (NRC). 1991a. *Technical Position on Waste Form, Revision 1, Final Waste Classification and Waste Form Technical Position Papers.* Washington DC: US NRC. January.

U.S. Nuclear Regulatory Commission (NRC). 1991b. *Technical Position on Waste Form, Revision 1, Appendix A, Cement Stabilization.* US NRC, Office of Nuclear Material Safety and Safeguards, Washington, DC. January.

US EPA. 1986a. *Test Methods for Evaluating Solid Waste.* EPA SW-846. 3rd edition. Office of Solid Waste and Emergency Response, Washington, DC.

US EPA. 1986b. *Land Disposal Restrictions.* 40 CFR 268, Fed. Reg. 51, 40638. November 7.

US EPA. 1988. *Field Studies of In-Situ Soil Washing.* EPA/600/S2-87/110. Cincinnati, OH. February.

US EPA. 1989a. *Immobilization Technology Seminar.* CERI-89-222. Cincinnati, OH. October.

US EPA. 1989b. *Preparing Perfect Project Plans.* EPA/600/9-89/087. Cincinnati, OH. October.

US EPA. 1989c. *Land Disposal Restrictions for Third Scheduled Wastes.* 54 FR 48439, N.

US EPA. 1990a. Superfund LDR guide #6B. Superfund publication: 9347.3-06BFS. Washington, DC. September.

US EPA. 1990b. *Toxicity Characteristic Leaching Procedure.* 40 CFR 261, Fed. Reg. 55, 11863. March 29.

US EPA. 1991a (draft). *Engineering Bulletin: Solidification/Stabilization of Inorganics and Organics.* Washington, DC. November.

US EPA. 1991b. *Engineering Bulletin: In-Situ Soil Flushing.* EPA/540/2-91/021. Cincinnati, OH. October.

US EPA. 1991c. *SITE Technology Demonstration Summary: International Waste Technologies/Geo-Con In Situ Stabilization/Solidification Update Report.* EPA/540/S5- 89/004a. Cincinnati, OH. October.

US EPA. 1992a. *Handbook on Vitrification Technologies for Treatment of Hazardous and Radioactive Waste.* EPA/625/R-92/002. Office of Research and Development, Washington, DC.

US EPA. 1992b. *Land Disposal Restrictions for Newly Listed Wastes and Hazardous Debris.* 40 CFR 268, 57 Fed. Reg. 37194. August 18.

US EPA. 1994a. *Geosafe Corporation In Situ Vitrification Technology.* SITE Technology Capsule. Cincinnati, OH.

US EPA. 1994b. *Land Disposal Restrictions Phase II — Universal Treatment Standards, and Treatment Standards for Organic Toxicity Characteristics Wastes and Newly Listed Wastes, Final Rule.* Washington, DC. September 19.

US EPA. 1995a. *Guide to Documenting Cost and Performance for Remediation Projects.* EPA-542-B-95-002. Washington, DC. March.

US EPA. 1995b. *Land Disposal Restrictions — Phase IV.* 40 CFR 148, 268, and 271, Fed. Reg.60, 43654. August 22.

US EPA. 1995c. Geosafe corporation in situ vitrification. *Innovative Technology Evaluation Report.* Washington, DC.

US EPA. 1995d. *Hazardous Waste· Identification and Listing.* 40 CFR 260, 261, 266, and 268. Fed. Reg. 60, 66344. December 21.

US EPA. 1996. *Land Disposal Restrictions Phase III.* 40 CFR 148, 268, 271, and 403. Fed. Reg. 61, 15566. April 8.

Van Dalen, A. and J.E. Rijpkema. 1989. *Modified Sulphur Cement: A Low Porosity Encapsulation Material for Low, Medium and Alpha Waste.* Nuclear Science and Technology. EUR 12303. Commission of the European Communities, Brussels.

Vectra Technologies, Inc. 1994. Manufacturers Product Literature.

Weitzman, L., L.E. Hamel ,and S.R. Cadmus. 1987. *Volatile Emissions from Stabilized Waste in Hazardous Landfills.* US EPA Contract 68-02-3993. Research Triangle Park, NC. August 28.

Werner and Pfleiderer Corp. 1976. *Topical Report Radwaste Volume Reduction and Solidification System.* Report No. WPC-VRS-1. Waldwick, NJ.

White, T.L., R.D. Peterson, and A.J. Johnson. 1986. Microwave processing of remote-handled transuranic wastes at Oak Ridge National Laboratory. *Waste Management '86 Proceedings.* R. G. Post and M. E. Wacks (eds.). pp 263-267.

Wilson, D.J. and A.N Clarke (eds.). 1994. *Hazardous Waste Site Soil Remediation* (Chapter 3). New York: Marcel Dekker, Inc.

WMX Technologies. 1993. TECHBytes, No. 4. OakBrook, IL. August.

Wright , S.L., R. Jones, J.F. McClelland, and P. Kalb. 1994. Preliminary tests of an integrated process monitor for polyethylene encapsulation of radioactive waste. *Stabilization and Solidification of Hazardous, Radioactive and Mixed Wastes.* ASTM STP 1240. T.M. Gilliam and C.C. Wiles (eds.). Philadelphia, PA: American Society for Testing and Materials.